HEALING WITH AMINO ACIDS

Survive Stress, Pain, Anxiety, and Depression

Without Drugs

What to Use and When

Billie J. Sahley, Ph.D.

Katherine M. Birkner, C.R.N.A., Ph.D.

Pain & Stress Publications®
San Antonio, Texas
1998

Note to Readers

This material is not intended to replace services of a physician, nor is it meant to encourage diagnosis and treatment of illness, disease, or other medical problems by the layman. This book should not be regarded as a substitute for professional medical treatment and while every care is taken to ensure the accuracy of the content, the authors and the publisher cannot accept legal responsibility for any problem arising out of experimentation with the methods described. Any application of the recommendations set forth in the following pages is at the reader's discretion and sole risk. If you are under a physician's care for any condition, he or she can advise you whether the programs described in this book are suitable for you.

This publication has been compiled through research resources at the **Pain & Stress Center**, San Antonio, Texas 78229.

1st Printing, June 1995
2nd printing, November 1995
3rd printing, March 1997
4th Printing, June 1998
Printed in U.S.A

A Pain & Stress Publication®
Vicki Worthington, Editor

Additional copies may be ordered from:
Pain & Stress Center
5282 Medical Drive, Suite 160, San Antonio, TX 78229-6023
1-800-669-2256

Library of Congress Catalog Card Number 98-65760
ISBN 1-889391-10-7

Dedicated . . .

To a new generation of physicians, therapists and educators who seek to find the natural alternative that frees the addicted from the prison of prescription drugs and gives them God's greatest gift—the freedom to make a choice.

To Robert Michael Benson, M.D., who taught me how to live my impossible dream.

To Julian Whitaker, M.D., one of God's gifted healers and an inspiration to all those who reach out.

To Candace Pert, Ph.D., neuroscientist, whose research opened many doors.

To a little boy named Scooter.

And to the Lord, for always lighting our paths.

Contents

Introduction

Orthomolecular therapy means supplying the cells with the right mixture of nutrients. Many diseases are known to be the result of the wrong balance of essential nutrients in the body. Adjusting the diet, eliminating junk foods, and ingesting the proper doses of essential vitamins, minerals, and amino acids, can correct the chemical imbalance of disease.

The orthomolecular approach helps patients become more aware of our dangerously polluted environment and nutrient-stripped refined foods. The orthomolecular approach is both corrective and preventive. Meganutrient therapy has become a part of orthomolecular medicine. While becoming widely recognized that orthomolecular therapy cures patients by correcting brain-chemical imbalances, it is little known that in certain combinations meganutrients can be as immediately effective as potent pain killers or tranquilizers. Meganutrients treat the whole person's biochemical imbalances; they can be of immediate and long-term benefit. The types of treatment offered by orthomolecular doctors and therapists vary, but the mainstream of their work focuses upon meganutrient therapy and diagnostic tests. Treatment with adequate nutrients distinguishes orthomolecular medicine from allopathic medicine.

Orthomolecular therapy considers every individual biochemically unique. Every patient has a very different nutrient and amino acid requirement. Application of this therapy balances each individual's need, and the mind and body join in a state of homeostasis—a condition where everything in the body is in balance and capable of resisting environmental changes, while regulating internal metabolic function.

Nutrition affects every tissue of the body. Under conditions of poor nutrition, the kidneys stop filtering, the stomach stops digesting, the adrenals stop secreting, and other organs follow suit.

Good nutrition is essential to the preservation of health and the prevention of disease. Meganutrient therapy has become a part of orthomolecular medicine. As it continually expands this therapy incorporates all our biological interactions with food, water, air, and light play an important part of good health and prevention of illness, if taken in proper amounts.

Special Information

The major focus of this book is amino acids. As you will notice, at times we discuss the use of herbs, vitamins, minerals, and other important nutrients. We included this important information to make you aware of all available options for maximum benefit. All the products described in this book are used at the Pain & Stress Center.

One of the major functions of the Pain & Stress Center is research. Our goal, through research, is to supply our readers, patients and customers the latest information to help you and your family achieve optimum health. If you have any questions regarding any of the products discussed in this book, call the Pain & Stress Center at 1-800-669-2256.

You will notice under some conditions such as Depression, we mention several products. This does not mean you need all of the products. You may find some products work better for you than others. Everyone's biochemically unique and responds to supplements in a different way. Our goal includes giving you all the information we have available to help you find the best and most effective resources for you and your family.

Neurotransmitters, Brain Language

Amino acids and brain function go hand in hand. Understanding your brain function gives you a more comprehensive picture of how to use various amino acids to effectively treatment of pain, stress, anxiety and depression. Your body needs and uses basic nutrients every day. These include vitamins, minerals, proteins, carbohydrates and fats. If you were to take the water and fat out of your body, 75% of the remainder would be protein. Muscles, cell membranes, enzymes, and the neurotransmitters are all proteins.

The brain controls every cell in the human body. Its commanding presence is responsible for all sensation, movement, thought, behavior, and a lifetime of memories and dreams. The importance of a healthy, well-nourished, efficiently functioning brain cannot be overstated. This three-pound powerpack comprises less than 2% of your total body weight. Your brain regulates your breathing, heartbeat, body temperature and hormone balance. Speech at any level would be impossible without needed nutrients for proper brain function.

Yet, in spite of the absolute importance of a smoothly functioning brain, it is the most poorly nourished organ in the human body. Ten billion neurons (brain cells) cry to be fed constantly. They need amino acids, vitamins, minerals, oxygen, and fatty acids. The neurons' needs must be satisfied every minute of every day of your entire life. All these nutrients are supplied to the brain via the bloodstream. If bloodflow to the brain is interrupted even for 20 seconds, unconsciousness will result. If this mighty powerpack is deprived of blood or oxygen for more than 7 to 10 minutes, it will die. The brain feeds on energy in the form of the chemical ATP, adenosine triphosphate. This energy fuels the neurotransmitters, the chemical language of the brain, conducts electrical impulses, transports proteins throughout the cells, extends new nerve connections to other brain cells, and rebuilds worn-out cell membranes. The brain must create its own energy for the billions of neurons that it must feed, and cannot borrow or steal this energy from other parts of the body.

How does the brain make this energy? Inside each neuron resides hundreds of little structures called *mitochondria.* These function as power plants for the cells. These little power plants burn fuel to generate the crucial ATP energy. The brain's very life depends on it. While virtually all organs and tissues of the human body can burn either fat or sugar for fuel, the brain can burn only sugar (glucose) under normal, nonfasting conditions. This glucose requirement creates potential problems for the brain. The brain cannot store sugar in its cells, so it totally depends on a second-by-second fuel delivery by the blood circulating through the brain. Brain cells use 50% of all glucose in the bloodstream for fuel and 20% of all inhaled oxygen. The brain's ability to claim this large amount of glucose depends upon a bloodstream relatively free of the blood-sugar-lowering hormone, insulin; thus, the importance of chromium picolinate in the diet, as it inhibits the release of insulin.

The way to ensure adequate glucose to the brain is to avoid simple sugar foods such as candy, pastries, drinks, etc. These high-sugar foods easily and powerfully trigger insulin release. Complex carbohydrate foods such as whole grains, vegetables, nuts, peas, beans, and seeds are much more desirable and a "brain-friendly" source of sugar. They are nature's *timed-release* sugar supplements.

Each of the brain's billions of neurons functions as a microcomputer. Inside each neuron, nerve impulses are conducted electrically. However, when information exchanges from one neuron to another, the brain uses chemicals called neurotransmitters to allow brain cells to communicate with each other (chemical language).

There are approximately fifty different neurotransmitters, but the communication conducted between brain cells uses only about ten major neurotransmitters. Certain neurotransmitters carry pain sensations, while others order voluntary muscle movement; some cause excitatory emotional responses, others are inhibitory. The neurotransmitters that govern our excitatory emotional responses are called catecholamines, noradrenaline (norepinephrine) and adrenaline (epinephrine), derive from the amino acids phenylalanine and tyrosine. Our reactions to everything we encounter, the way we are stirred by a song or an old picture, angered by an argument or emotional pain inflicted by someone we love, or amused by something we see on television—all depend on the chemical language of the brain; specifically, neurotransmitters. Too much or too little of any of these substances will make us under or overreact according to the stimulus.

How we feed the brain directly affects our production of neurotransmitters. Neurotransmitters determine our mental and emotional state of well-being. Proper nutrition and supplementation can correct or enhance mind, mood, memory, and behavior.

No drug currently in wide use, medical or recreational, addresses the root of neurotransmitter problems. Drugs merely stimulate temporary excessive release of preexisting neurotransmitter stores. They do not increase production of neurotransmitters. This fact explains why drugs often lose their effect over time, with chronic use. Once preexisting neurotransmitter stores are exhausted, the drug is unable to stimulate further neurotransmitter release into the synapses between neurons. Hence, the phenomenon known as drug-abuse "crash" frequently occurs.

The Synapse

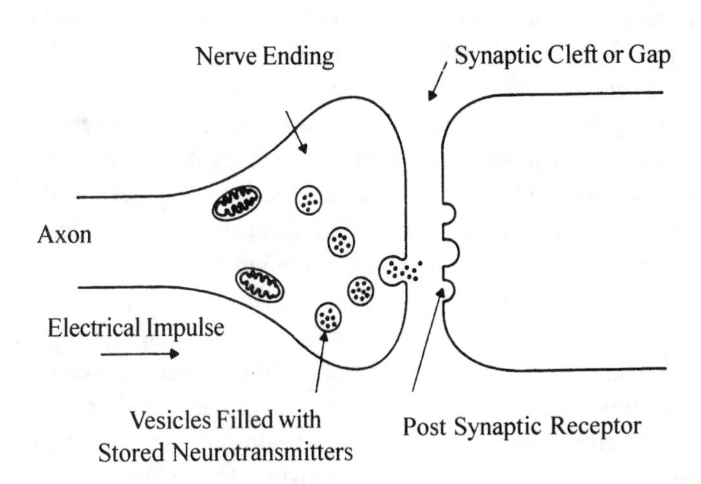

In the brain, neurotransmitters like serotonin and norepinephrine carry signals from one nerve to the next across the gap (synapse) between the two. Tryptophan enters the presynaptic nerve cell, where it converts to serotonin. As more Tryptophan enters the cell, more serotonin gets released into the synapse. Antidepressant drugs work by keeping more neurotransmitters in the synapse.

Greater transport into the brain of the relevant amino acids, vitamins, and minerals augment nourishment of the brain when there are less-than-adequate neurotransmitter levels. And this, in turn, requires higher levels of amino acids, vitamins, and minerals. If levels of these nutrients in the typical American junk-food diet truly and adequately promoted optimal neurotransmitter levels in the brain, we would not be seeing the epidemic level in the U.S. of antidepressant, anti-anxiety, anti-manic, anti-schizophrenic, and recreational drug use. Virtually all these drugs act either by increasing the synaptic release of brain cell neurotransmitters, without increasing their production, or by "pinch-hitting" for neurotransmitters whose synaptic levels are chronically low.

All substances of abuse either raise or lower consciousness, and deplete the available neurotransmitters needed to prevent or alleviate depressed moods. When you use drugs escape is the primary goal, you cannot escape stress, anxiety, depression, or grief. You merely prolong the healing process. Prescription drugs for stress do not restore or resolve, they *merely use available neurotransmitters.*

All major neurotransmitters are made from amino acids and from dietary protein. One of the dangers of a low-protein diet includes not producing enough amino acids to make adequate brain neurotransmitters. Apathy, lethargy, difficulty concentrating, loss of interest, and insomnia all result from not enough amino acids in the diet. A.D.D. and hyperactive children, as well as adults, have low levels of neurotransmitters. Drug use does not produce or increase production of neurotransmitters. Drugs *only* address symptoms.

Children, as well as adults, on prolonged drug use or alcohol, have dangerously low levels of neurotransmitters. They display panic and anxiety because of the deficiency of neurotransmitters. Once proper supplementation is achieved, the symptoms of panic and anxiety decrease noticeably. You cannot restore the brain chemistry overnight by megadosing. Deficiencies must be established, then adequate amounts of amino acids, vitamins, and minerals must be implemented. All this is part of the healing process. You have taken the first step by purchasing this book.

Blood-Brain Barrier

The Protective Barrier

In 1968 Dr. Linus Pauling published an article in *Science* magazine describing the unique function of the "blood-brain barrier." Dr. Pauling postulated the brain's disadvantage in acquiring the high levels of vitamins and minerals it needs is due to the presence in the brain of a unique blood brain barrier (B.B.B.). The B.B.B.'s major purpose is to protect the brain from water-soluble toxins. Ironically, however, most of the major brain nutrients—glucose, vitamins B and C, minerals, and amino acids—are all water soluble. Therefore, the B.B.B. makes it difficult for the brain to absorb the large nutrients needed for delivery to other organs and tissues.

Scientific studies reported by the National Institute of Health (NIH) reveal mild nutrient deficiencies typically cause changes in mood and behavior. Other memory and mental abnormalities will present early detectable signs of nutrient deficiency. Dr. Pauling recommended an increase in the daily intake of B6, Vitamin C, and magnesium. This is far beyond R.D.A. (recommended daily allowance) levels as this produces higher blood levels of these nutrients, thus increasing their penetration of the blood-brain barrier.

Research done at M.I.T. as early as 1970 pointed to a conclusion that brain neurotransmitter levels were controlled totally by the brain, itself, independent of dietary intake of various amino acids. Further research at M.I.T. pointed to brain neurotransmitter levels are significantly influenced by a single meal. A meal rich in protein will encourage high adrenaline/norepinephrine levels with consequent high alertness and assertiveness. A meal rich in simple sugars and low in protein will usually increase brain serotonin levels. This leads to a very relaxed or even sleepy state.

Alan Gelenberg, M.D., Department of Psychiatry at Harvard Medical School, found tyrosine to be more effective than antidepressants. Dr. Gelenberg noted that those with stress burnout and overload responded extremely well to tyrosine and B6. Tyrosine is the amino acid precursor of noradrenaline, adrenaline and dopamine.

Glutamine, one of the most plentiful amino acids in the brain, provides a major alternative fuel source for the brain with low blood-sugar levels. The amino acid, glycine, also helps alleviate sugar cravings, or that low feeling in mid afternoon. Glutamine is important as the precursor for GABA (Gamma Amino Butyric Acid). GABA and glutamine help those with problems of concentration and A.D.D. Glutamine also helps the brain dispose of waste ammonia, a protein breakdown byproduct irritating to the brain cells even at low levels.

Adequate vitamin and mineral intake also promotes optimum brain neurotransmitter production. B6 is essential to convert all amino acids. Pyridoxal 5' Phosphate (P5'P), the active or coenzyme form of Vitamin B6, is essential to carbohydrate, fat, and especially, protein metabolism. P5'P can be used in place of B6, is safe for children, and has no adverse side effects. The importance in providing optimal brain nutrition to facilitate optimal brain neurotransmitter production cannot be overemphasized.

Neurotransmitters are produced within neurons and stored in small packets, called *vesicles,* near the end of the neurons. People are unaware of their brain's activity. The limbic system, the feelings part of the brain, stores all fears of the past, present, and future. The GABA -receptor sites control the firing of anxiety-related messages at the cortex, the decision-making part of the brain. The brain power plant *never shut down and must be fed constantly.* When an electrical current flashes down the length of a neuron, neurotransmitter molecules secrete into microscopic synaptic gaps between two adjacent neurons. Once secreted into the synaptic gap, or synapse between two neurons, the enzymes in the synapse either destroy the neurotransmitters' molecules, or recycle back into the preceding neuron.

Amino Acids for Brain and Body Function

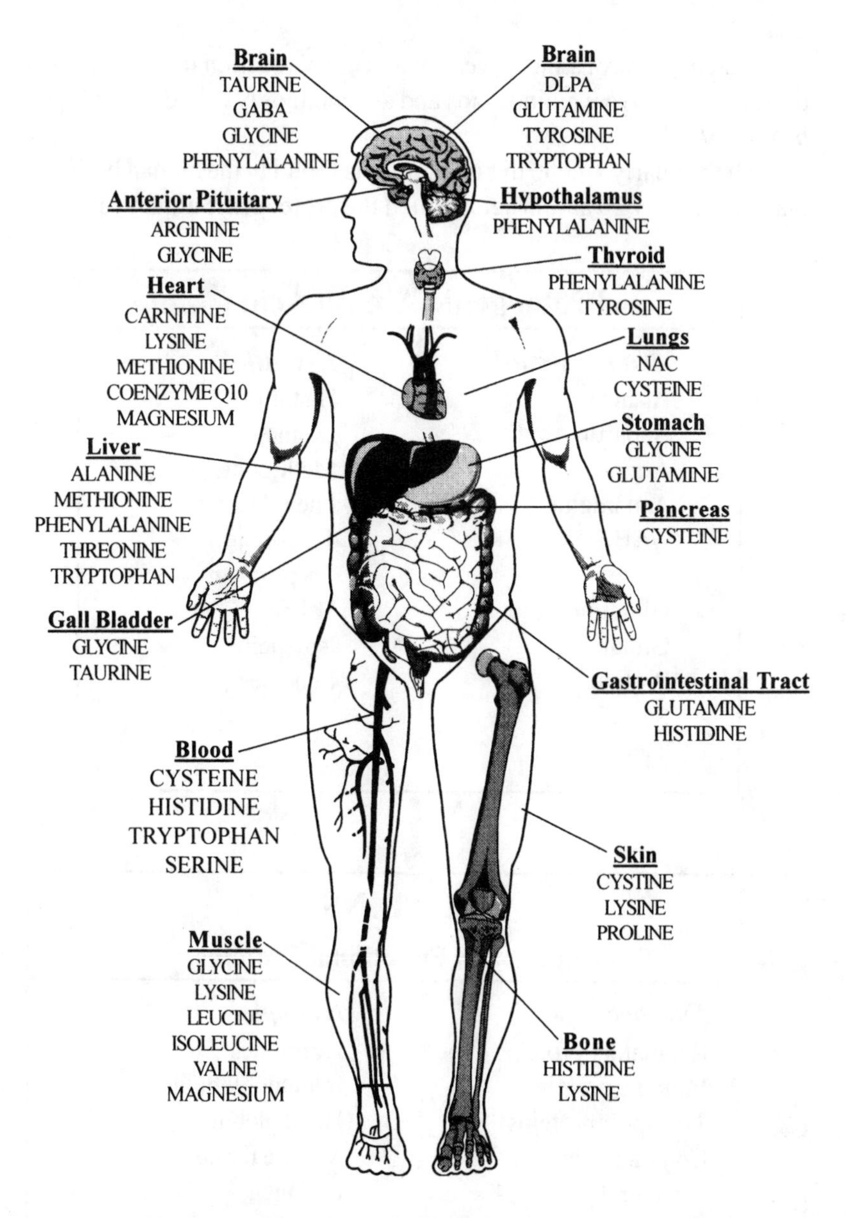

Always add magnesium and B6 or P5'P to all amino acids.

Understanding Amino Acids/ Proteins

Proteins control almost every biochemical reaction in the body. All proteins derive from amino acids and are commonly called the *building blocks of life.*

All the nearly 40,000 distinct proteins found in the human body are made from only 20 amino acids called the proteogenic amino acids.

Proteogenic Amino Acids	
Nonessential	***Essential***
Alanine	Histidine
Aspartic Acid	Lysine
Arginine	Methionine
Asparagine	Phenylalanine
GABA	Threonine
Cysteine	Tryptophan
Glutamic Acid	Valine
Glutamine	Isoleucine
Proline	Leucine
Serine	
Tyrosine	

PROTEINS Divide into Six Functional Proteins	
Function	***Example***
Regulatory proteins	Hormones
Immune proteins	Immunoglobulins
Transport proteins	Hemoglobin
Contractile proteins	Muscle tissue
Structural proteins	Collagen
Enzymes	Proteinase

Source: *Interpretation Guide To Amino Acid Metabolism and Analysis,* Aatron Medical Services

Amino acids can be broken down into essential and nonessential categories. The body cannot synthesize essential amino acids; they *must* be obtained from the diet. Nonessentials can be synthesized in the body, and are not mandatory in the diet. Although the body can manufacture nonessential amino acids, an abnormality in the production of the nonessential amino acids can be detrimental metabolically. Some become conditionally essential amino acids under certain circumstances; i.e., infancy, illness, stress, etc.

Most of the protein in the body resides in the skeletal muscles. Only 0.1% of all protein is found as free amino acids. Plasma amino acids represent what is available to the body at the current time. A deficiency in one of the proteogenic amino acids can limit the body's ability to make an optimal number of certain proteins. Deficiency effects result in both health and disease.

Classification of Amino Acids

Nonessential	***Conditionally Essential***	***Essential***
Alanine	Arginine	Histidine
Asparagine	Cysteine	Isoleucine
Aspartic Acid	Cystine	Leucine
Glutamic Acid	Glutamine	Lysine
Glycine	Taurine	Methionine
Proline	Tyrosine	Phenylalanine
Serine		Threonine
		Tryptophan
		Valine

Neurotransmitters & Neuropeptides

Dr. Candace Pert is Professor of Molecular and Behavioral Science at George Washington University and formerly Chief of the section on brain biochemistry and Clinical Neuroscience at National Institute of Mental Health. Dr. Pert discovered the opiate, GABA, and many other peptide receptors in the brain and body. Her discoveries lead to an understanding of the chemicals that travel between the mind and body, such as neurotransmitters.

Everything in your body is being run by messenger molecules called neuropeptides or neurotransmitters. A peptide is made up of amino acids, the building blocks of proteins. Peptides are amino acids strung together very much like pearls strung in a necklace. Peptides are found throughout the brain and body. They are extremely important because they mediate intercellular communication throughout the brain and body. Dr. Pert calls neuropeptides and their receptors "the biochemical correlates of emotion."

Within the brain itself are about sixty neuropeptides. Peptides are found in the parts of the brain that mediate emotion. Neuropeptides allow the systems of the body to talk to each other. They control the opening and closing of the blood vessels in your face and throughout the body. Neuropeptides carry messages within the brain, and from the brain to all the body. Neuropeptides direct energy in the body to where it is needed the most. During times of high stress, energy must be directed not only to the brain, but to those parts of the body directly affected by the stress.

One reason the amino acid, GABA, so effectively relieves stress and anxiety is because it is able to reach the GABA receptors throughout the body and brain. Dr. Pert's research established two realms of emotion—physical and mental, found in every cell of your body. In her book, *Molecules of Emotion,* Dr. Pert states "Emotions are at the nexus between matter and mind, going back and forth between the two and influencing both." The molecular bases of our emotions are inseparable from our physiology. Part of being a healthy person means being well integrated and at peace, with all systems acting together. Neuropeptides and neurotransmitters are the keys to understanding our fear, anxiety, depression, and every emotion we feel.

The study of GABA, other amino acids,and how they affect the brain and behavior is making substantial contributions to the understanding of disease in man. Research done in the field of psychoneuroimmunology confirms disease can originate from within individuals. Major contributing factors include continuous stress overload, the environment, nutrient imbalances, and changes in brain chemistry. The immune system directly effects stress and the quality of life. The different functions of amino acids and how they affect the brain and behavior provide a major focal project for scientist and researchers.

A new age of medicine has emerged, and incorporates the substantial evidence that nutrient deficiencies can and do influence mind, mood, memory, and behavior. Amino acid requirements in the body and brain are tremendously increased by disease and inborn metabolic errors. Any time a person undergoes prolonged periods of stress, anxiety, depression, or grief, they require more amino acids; some more than others. The reason for the different requirements: biochemical individuality.

Every individual has a distinct chemical composition. The brain, glands, and bones are distinct for each person, not only in anatomy, but also in chemical composition. This does not mean that chemical compositions are fixed throughout life, or that they are not influenced by nutrition. Nutrition and amino acid deficiencies affect every tissue in the body, and most importantly the brain.

Limbic System & Mental Distress Pathways

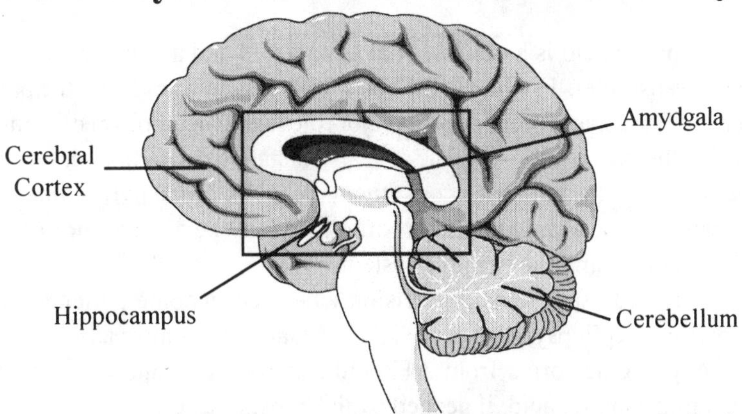

The limbic system is the region of the brain where emotion and moods are regulated and conveyed to the cerebral cortex. The limbic system contains the inhibitory neurotransmitters, GABA, glutamine, glycine, and serotonin that modulates anxiety messages in the brain.

Understanding Amino Acids

Alanine

Alanine is a nonessential amino acid, and functions as an inhibitory neurotransmitter in the brain. Highest concentrations of alanine are found in the muscles. During hypoglycemia, alanine may provide an alternative source of glucose.

Elevated alanine levels in the blood can cause drug-resistant seizure disorders or severe depression. Low alanine levels are often seen with low glycine and taurine, and when the Branch Chain Amino Acids (BCAAs) are deficient. Normal alanine metabolism requires the presence of B6. Alanine is essential for the normal metabolism of tryptophan.

Best food sources for alanine include wheat germ, turkey, duck, cottage cheese, and sausage. Usual supplemental dose range is 200 to 600 mg, daily.

Asparagine and Aspartic Acid

Aspartic acid is a nonessential amino acid and a major excitatory neurotransmitter. Formed in the body from glutamic acid with the presence of B6, aspartic acid plays a major role in the metabolism of ammonia via the urea cycle. Aspartic acid also metabolizes carbohydrates via the Krebs cycle, and forms constituents of DNA called pyrimidine and orotates. Research indicates aspartic acid may be a stimulator of the thymus gland and the immune system.

Elevated aspartic acid levels may be seen in some patients with depression, epilepsy, stroke, high BCAAs, and low ornithine.

Asparagine forms from ATP and aspartic acid, and can convert back into aspartic acid, if needed by the body.

Good food sources of aspartic acid include meats such as pork, turkey, sausage, chicken, wild game, wheat germ, cottage and ricotta cheeses.

Branched Chain Amino Acids (BCAA)

Leucine, Isoleucine, and Valine

The branched chain amino acids include leucine, isoleucine, and valine. As essential amino acids, BCAA's must be obtained from foods. They are especially involved in stress reactions, energy, and muscle metabolism. BCAAs are unique because the skeletal muscles use them directly as an energy source, and they promote protein synthesis.

The BCAAs are similar structurally, but have different metabolic routes. Leucine solely goes to fats; valine solely to carbohydrates; and isoleucine to both. A valine deficiency appears as neurological defects in the brain. Muscle tremors mark an isoleucine deficiency. Stress states such as infections, trauma, surgery, fever, cirrhosis, and starvation, require proportionally more leucine than valine or isoleucine. Diseases such as hepatitis, cirrhosis, hepatic coma, or liver disease, lower the levels of BCAAs. BCAAs, as well as other amino acids, are commonly fed intravenously to chronically ill patients. The BCAAs, particularly leucine, stimulate protein synthesis, increase the reutilization of other amino acids in many organs, and decrease protein breakdown.

As stress rises, total caloric intake needs increase, primarily due to increased protein requirements. Stress causes proteins to break down rapidly, and increases amino acid utilization three to four fold. About 30% of the diet should be protein or amino acids, especially when the body undergoes severe stress. But when taken in supplement form, BCAAs decrease the rate of amino acid and protein breakdown. More BCAAs and B6, or P5'P (pyridoxal 5' phosphate), are needed as stress or disease accelerates.

Utilization of the BCAAs by athletes, especially weight lifters, increases available energy. BCAA helps replace steroids used by those who want to build muscle mass. The BCAAs, especially leucine, greatly produce energy under many kinds of stress—from trauma, surgery, fever, infection, muscle training, and weight lifting. With prolonged exercise, about 5 to 10% of the energy used comes from amino acids, especially BCAAs. BCAAs should be used in *all* stress situations. The amount will depend on your physical state and stress level. Normal dosage of BCAA is 1,000 to 3,000 mg per day, divided. BCAAs should be taken together and not singularly. The ingestion of only one BCAA, particularly leucine, decreases the plasma tissue levels of valine and isoleucine.

Carnitine

Carnitine was discovered in 1905 from extracts of meats, but no physiological role for carnitine could be found until fifty years later. Early research indicated carnitine to be essential to the diet, but later it research discovered the body produced carnitine from lysine and methionine provided sufficient amounts of niacin, Vitamin B6, C, and iron were present. If a carnitine deficiency exists, deficiencies of lysine and methionine also exist.

Carnitine is a nonessential amino acid synthesized from lysine and methionine in the liver, kidney, and brain. Concentrations of carnitine are forty times greater in the muscles than in the plasma. Major sources of carnitine in the diet are meat, especially organ meats such as the liver, and dairy products. Vegetables, grains, and fruits contain little or no carnitine. Vegetarians are therefore more susceptible to deficiencies of carnitine, but also of lysine and methionine.

In 1973 research showed carnitine deficiencies exist in some people for various reasons. Between 1980 and 1983, almost 300 studies were published investigating carnitine's nutritional value and anomalies of carnitine metabolism causing clinical symptoms. Some carnitine deficiency symptoms include impaired lipid (fat) metabolism; lipid accumulation in the skeletal muscles, heart muscle, and liver; and progressive muscle weakness with a buildup of fats in the muscle cells. In children carnitine, deficiency may manifest as loss of muscle tone, failure to thrive, swelling in the brain, recurrent infections, hypoglycemia, and heart disturbances.

Carnitine is essential to the transportation of long-chain fatty acids into the cells where the fats can be converted to energy. Recent research indicates carnitine plays an important role in converting stored body fat into energy, energizing the heart, reducing angina attacks, controlling hypoglycemia, and is beneficial with diabetes, liver disease and kidney disease. Carnitine primarily regulates fat burning in the body. Carnitine transports large fat molecules into the part of the cells where fats can be converted into energy. If your level of Vitamin C is low, you can have an apparent deficiency of carnitine. If carnitine is absent or deficient, many fats cannot be burned. The fats build up within the cell and bloodstream as triglycerides and cholesterol. Carnitine supplementation significantly reduces serum triglycerides and cholesterol levels, while increasing HDL (high density lipids or good cholesterol). Dosages are 1,000 to 3,000 mg per day, divided.

Normal cardiac function is dependent on adequate amounts of carnitine. In angina, carnitine helps improve oxygen utilization and energy metabolism by the heart muscle. By improving the fat utilization and energy production, carnitine prevents the buildup of toxic fatty acid metabolites. These metabolites can cause cell-membrane damage throughout the heart, contributing to impaired heart-muscle contraction, arrhythmias (irregular beats), and death of the heart muscle. Carnitine supplementation helps prevent the production of these fatty acid metabolites.

Many clinical trials have shown that carnitine improves angina and heart disease, and is beneficial in recovery from a heart attack, arrhythmias, and congestive heart failure. Oral supplementation with carnitine helps normalize heart carnitine levels, thereby allowing the heart to use its limited oxygen supply more efficiently. In one study, patients released from the hospital received 4 grams of carnitine daily, and recovered more quickly from their heart attacks. They showed dramatic improvements in heart rate, blood pressure, rhythm disturbances, angina attacks, and clinical signs of impaired heart function compared to the control group.

Carnitine increases heart rate, pressure rate, heart-pacing duration, and decreases left ventricular and diastolic pressures. It increases muscle strength, and is beneficial to heart patients by increasing exercise endurance at doses of 20 to 40 mg/kg (1,400 to 2,800 mg/day). Other studies indicate that carnitine lowers the exercise heart rate, and extends the time of exercise prior to the onset of angina at doses of 40 mg/kg (about 2,800 mg in 150 pound individuals). Carnitine proves very beneficial in congestive heart patients.

Intermittent claudication is intermittent pain in the calf muscle. This is often described as a cramp or tightness causing pain in the calf, much like angina occurs in the heart. Carnitine improves distances walked without pain in intermittent claudication and other peripheral vascular diseases. In one study, 2 grams of carnitine, twice daily, showed a 75% increase in the walking distance after three weeks of supplementation. This results from the improved energy metabolism within the muscle.

In patients with uremia or kidney disease, carnitine may reduce the risk factor for atherosclerosis and coronary heart disease. Carnitine dramatically reduces triglycerides and cholesterol levels, while increasing HDL (good cholesterol) levels. In addition, carnitine reduces muscle cramping and muscle weakness by restoring muscle carnitine levels. Carnitine deficiencies usually exist in hemodialysis patients due to decreased carnitine synthesis.

Diabetics have reduced blood levels of carnitine. Since cardiovascular disease and reduced kidney and liver function are found in diabetics, supplementing with carnitine is encouraged.

Reduced levels of carnitine in skeletal muscles are seen in patients with various muscular dystrophies. The muscular weaknesses experienced by these patients is believed to result from the low carnitine levels. Carnitine supplementation increases muscle strength, while decreasing lipid levels.

Carnitine also plays an important role in the production of heat in brown fat. The "brown fat" helps us acclimate to cold temperatures, and is thought to help determine how much of the food eaten is burned for heat, and how much is stored as body fat. During poor-weight loss diets, carnitine prevents the accumulation of ketones in the blood stream. Ketosis can be life threatening, if uncontrolled. And ketonic states can cause the loss of potassium, magnesium, and calcium.

In cancer patients receiving chemotherapy, carnitine has been shown to protect the heart during adriamycin chemotherapy. Additionally, carnitine helps alleviate some of the toxicity associated with chemotherapy through its effect on the mitochondrial integrity.

Carnitine helps improve the symptoms attributed to anticonvulsant medications such as Valproic acid (Depakote, Depakene) and carbamezepine (Tegretol and Epitol). Carnitine may offer protection against drug toxicity.

Recent studies indicate carnitine may be a factor with depression. Supplementation with 1,000 mg per day, divided, is helpful.

But carnitine supplementation does not need to be limited to people with cardiovascular problems. Effective utilization of fatty acids by the heart and skeletal muscles depends upon ample supplies of carnitine. Supplementation with 1 to 2 grams, two to three times daily, results in significant improvements in cardiovascular function in response to exercise, especially endurance related activities. It may benefit healthy people as well as athletes by helping burn certain amino acids, called BCAAs or branch chain amino acids in the skeletal muscles. During fasting or prolonged strenuous exercise, the BCAAs provide a significant amount of metabolic energy. Ketones result from the incomplete burning of fat, and are extremely toxic to the nervous system and brain. Carnitine helps to significantly lower the ketone levels.

The daily dose of carnitine ranges between 1,000 to 4,000 mg in divided doses. Carnitine is very safe with no significant side effects ever reported in any human studies. But use *only* the L-carnitine form of carnitine. There are no know adverse interactions between carnitine and any drug or nutrient. When carnitine and coenzyme Q10 are combined, they appear to work synergistically.

Cysteine and Cystine

Cysteine is the considered conditionally essential, and is one of sulfur-containing amino acids. Cystine is the stable form of cysteine. Conversion of one to the other occurs in the body, as either is needed. If Vitamin C is not present, cysteine converts to cystine. Research now focuses on the protective role of dietary amino acids. Cysteine builds proteins, such as those in hair, and also helps destroy harmful chemicals in the body such as acetaldehyde and free radicals produced by smoking and drinking.

Cysteine spares methionine (another important amino acid) and can completely replace dietary methionine if the diet is supplemented by appropriate amounts of folic acid and Vitamin B12. Cysteine resides abundantly in proteins such as keratin in hair (12%) and trypsinogen (10%). Cysteine acts is a detoxifier. Heavy metals such as mercury, lead, and cadmium will be "tied up" by cysteine so it can be removed from the body.

Cysteine may protect heavy drinkers and smokers against acetaldehyde poisoning from chronic alcohol intake or smoking, according to Dr. Herbert Sprince at the V.A. Hospital in Coatesville, Pennsylvania, and Thomas Jefferson University in Philadelphia.

Pearson reports cysteine effective "not only in preventing hangovers, but also in preventing brain and liver damage from alcohol, and in preventing damage such as emphysema and cancer caused by smoking." Cysteine has been found to offer a degree of protection against radiation.

Recently, Dr. William Philpott postulated that cysteine is necessary for the utilization of Vitamin B6. His studies suggest "a majority of chronic degenerative illnesses, whether physical or mental, have a Vitamin B6 utilization disorder. The culprit causing this Vitamin B-6 utilization problem seems to be cysteine deficiency." Dr. Philpott recommends patients having the Vitamin B6 utilization problem take 1.5 grams of L-cysteine three times a day for a month, and then reduce it to twice a day. Vitamin C should always be taken with cysteine.

A specialized cysteine known as NAC, or n-Acetyl Cysteine, was first produced by Mead Johnson for the treatment of excess mucus. The primary use for NAC is as a mucus-reducing agent. Mucus strands are broken up to decrease viscosity and congestion associated with excess mucus and sinus drainage. Oral supplementation of NAC is used with allergies, bronchitis, chronic sinusitis, asthma, pneumonia, cystic fibrosis, and even the common cold. NAC modulates the production of, and the precursor to, glutathione. In turn, glutathione helps protect the body against natural and man-made oxidants. Glutathione is an antitoxin and a neurotransmitter. Supplemental NAC dosage is usually 600 mg, twice to three times daily.

Diabetics should not use cysteine because it can change glucose levels to change.

GABA
An Inhibitory Neurotransmitter

GABA (Gamma Amino Butyric Acid), an inhibitory neurotransmitter, is found throughout the central nervous system (CNS). GABA assumes an ever-enlarging role as a significant influence on pain, stress, anxiety, and depression as well as stress-induced illnesses. By January 1998, there were over 3,000 documents and texts on GABA, describing how it affects anxiety/stress in the brain.

If you examine a step-by-step process of what happens in the brain when you feel stress and anxiety, you would see how GABA works to slow down messages. Panic, anxiety, or stress-related messages begin to release numerous signals, and concurrently a physiological response begins to take place—the fight-or-flight syndrome.

The unceasing alert signals from the limbic system eventually overwhelm the cortex (the decision-making part of the brain), and the ability of the cortex and the rest of the stress network becomes exhausted. The balance between the limbic system, and in fact, the rest of the brain to communicate in an orderly manner depends critically on inhibition. GABA inhibits the cells from firing, diminishing the anxiety-related messages from reaching the cortex.

GABA fills certain receptor sites in the brain and body. This slows down and blocks the excitatory levels of the brain cells that are about to receive the anxiety-related, incoming message. When the message is received by the cortex, it does not overwhelm you with anxiety, panic or pain. You are able to maintain control and remain calm. But, if you are under prolonged stress or anxiety, your brain exhausts all the available GABA and other inhibitory neurotransmitters, thus allowing anxiety, fear, panic and pain to attack you from every direction. Your ability to reason diminishes. In a full blown anxiety or panic attack, physical symptoms include excessive sweating, trembling, muscle tension, weakness, loss of control, disorientation, difficulty breathing, constant fear, headaches, diarrhea, depression, and unsteady legs.

Research done at The Pain & Stress Center, in San Antonio, with patients suffering from all types of stress, pain, muscle spasms or anxiety/panic attacks, demonstrated pure GABA, 750 mg, can mimic the tranquilizing effects of Valium or Librium *without* the possibility of addiction or fear of being sedated. GABA fills the receptor in the brain and nourishes the brain with what should be there. Pure GABA dissolves in water, and is tasteless and odorless. The calming effect usually occurs within 10 to 12 minutes.

Tranquilizers provide only temporary relief. We have seen many patients on Xanax that still experience anxiety. They have been told it is not addicting—it is! *THERE IS NO SUCH THING AS A TRANQUILIZER DEFICIENCY!* Nutrient deficiencies can and do change behavior. Human behavior involves the functioning of the whole nervous system, and the nervous system requires amino acids. GABA, glutamine, and glycine proves vital for energy and the smooth running of brain functions.

B6 (pyridoxine) is GABA's most important partner. We have successfully used GABA, glutamine, and glycine with patients to ease anxiety, muscle pain/spasms, and nervous stomachs. GABA 750 is freeform, not combined with anything else. There is a GABA with niacinamide and inositol on the market, but let me caution you—*do not megadose*, with this form. If you do, you will have side effects. Side effects include tingling lips and extremities, rapid heart beat, shortness of breath, flushing, nausea, and increased anxiety. If this occurs, drink 16 ounces of water, and eat a couple of soda crackers.

Special Note: Magnesium is the stress mineral, and is involved in over 300 enzyme reactions in the body. Take magnesium along with B6. The best form for maximum absorption and tolerance is magnesium chloride (Mag Link). Magnesium chloride is the same form of magnesium present in the body.

Glutamic Acid

Glutamic acid is a nonessential amino acid which can be synthesized by the body, or be converted into glutamine and GABA. Glutamic acid or glutamate is thought to be an excitatory neurotransmitter. It acts as a brain ammonia detoxifier.

Since glutamic acid can be manufactured from aspartic acid, ornithine, arginine, proline, and alpha-ketoglutarate, no deficiencies of glutamic acid have ever been seen. Elevation of glutamic acid may be present in some schizophrenics, epileptics, and patients with gout. In fact, epileptics generally have an elevation of glutamic and aspartic acids, and have low levels of GABA, taurine, and glycine.

Rich food sources of glutamic acid include sausage, ham, bacon, yogurt, turkey, chicken, duck, cottage cheese, wheat germ, and granola.

Glutamine

Glutamine, a nonessential amino acid, is the third most abundant amino acid in the blood and brain. Glutamine, an inhibitory neurotransmitter acts as a precursor for GABA, the antianxiety amino acid. It helps the brain dispose of waste ammonia, a protein breakdown byproduct. Ammonia irritates brain cells, even at low levels. Recent scientific research demonstrates its link to the most important functions of the body's vital organs and musculoskeletal system. Glutamine assists the body in muscle development when illness causes muscle wasting—sometimes seen following a high fever, chronic stress, illness, or a traumatic accident. Glutamine provides a major alternative fuel source for the brain with low blood sugar levels.

In 1980 glutamine was reestablished as a conditionally essential amino acid; prior to 1980 glutamine was considered a nonessential amino acid. A conditionally essential amino acid means that under normal circumstances, the body can make (synthesize) adequate quantities of that amino acid; but in times of stress such as fever, illness, trauma, dieting, or chemotherapy, the body cannot make as much as it requires. An additional amount of the amino acid must be taken in nutrient form to prevent a deficiency.

Glutamine's most important function is strengthening the immune system. Glutamine supports the multiplication of selected white cells which strengthen the body's defense system. Glutamine aids other immune cells in killing bacteria, healing wounds, and maintaining and supporting glutathione, as an important antioxidant. Glutamine also supports pancreatic growth.

Scientists at NIH in 1970 found glutamine to be the most important nutrient for the intestinal tract. During times of illness, the body uses more glutamine to help tissue repair in the kidneys, intestines, and liver. For many years glutamine was considered a nonessential amino acid, but research over the past several years brought forth a wave of new, important information to change this view.

Every day researcher conduct more studies on the healing power of amino acids. Glutamine deserves special attention. Studies show glutamine supplementation during cancer treatment increased the effectiveness of many chemotherapy drugs, and increased tumor kill-off. Systemic infections (sepsis) decreased by up to 80% with glutamine supplementation. The mechanisms appear to be increased intestinal integrity and reduced intestinal ulcerations. Glutamine decreased weight loss and increased nitrogen balance. Following radiation, glutamine decreased mortality to the point of ameliorating toxicity. Additionally, glutamine augmented healing of radiated intestines.

Some amino acids help the body resist the effects of the radiation which is becoming a significant pollutant and potentially a worldwide problem. In cancer patients, glutamine enhances the effectiveness of chemotherapy and radiation treatments, while reducing the toxicity and damage to the body. Dosages vary in amounts, but a rule of thumb is 0.5 grams/kilogram of body weight daily. For best results, take glutamine prior to treatment and continue throughout therapy. Research has shown glutamine is the second-most important fuel for the cells' lining in the colon.

The main nutrient needed for intestinal repair is glutamine. Leaky-gut syndrome is recognized more often today due to the increased use of anti-inflammatory medications such as Motrin, Advil, Ibuprofen, Dolobid, Anaprox, Orudis, Naprosyn, etc. Leaky-gut syndrome makes the intestines more permeable and allows substances and foods which do not normally pass into the circulation to cross. Food allergies can result, causing more discomfort and pain. But glutamine helps the gut to heal and makes the intestines less permeable. Japanese researchers found glutamine helps stomach ulcers heal.

Glutamine helps clear the body of waste through the kidneys and liver. For those with impending surgery, glutamine supplements should be considered before, during, and after surgery. New research demonstrates the body's release of amino acids during times of stress includes one-third in the form of glutamine. Further research revealed the muscles synthesize glutamine as they break down in times of heavy stress. When glutamine was taken with balanced amino acids, muscle breakdown (atrophy) was essentially prevented.

Note: Over the past years there has been confusion regarding glutamine, glutamic acid, and glutamate. Glutamine is not glutamic acid, glutamate, MSG, or glutathione. Glutamine, GABA, and glycine are rapidly becoming the most important therapeutic amino acids of the twenty-first century.

Neurotransmitters in Brain Function

The amino acid trio of glutamine, GABA, and glycine, along with Vitamin B6, the cofactor, represent the major inhibitory neurotransmitters in the brain. Glutamine is found in the nerves of the hippocampus—the memory center of the brain, in the cranial nerves, and in many other areas of the brain. These three amino acids work together, and are inhibitory neurotransmitters. Anyone taking amino acids must take Vitamin B6 to metabolize these amino acids.

Glutamine studies reported intellectually impaired children and adults demonstrated an increase in IQ after taking glutamine in combination with ginkgo and Vitamin B6. Research done by Dr. Roger Williams at the University of Texas, Clayton Foundation, showed children and adults classified A.D.H.D. showed a marked improvement when taking glutamine, 250 mg to 1,000 mg, daily. Dosage depends on age and weight. At the Pain & Stress Center we use a Balanced Neurotransmitter Complex plus GABA, along with AC (Anxiety Control 24). If needed, extra glutamine powder and ginkgo are added. Results have been excellent. The Balanced Neurotransmitter Complex formula assists brain communication and allows the brain cells to talk to each other. Recent discoveries found 50 or 60 neuropeptides in the immune system, as well as in the brain. Each unique neuropeptide has its own receptor. These intercellular neuropeptides and receptors mediate communication among the brain, glands, and immune system. Neuropeptides are peptides made up of amino acids, the building blocks of proteins. Neuropeptides and their receptors form the biochemical correlates of emotion.

GABA and glutamine are not found *only* in the brain, but in the receptor sites throughout the body. Amino acids can and do change mind, mood, memory, and behavior. A particular herb, Ginkgo Biloba demonstates excellent results in enhancing concentration. Ginkgo increases blood flow to the head, and improves mental functioning and the ability to focus for longer periods of time. Ginkgo has also been helpful after a stroke by increasing circulation to the brain. In his book *Herbal Tonic Therapies,* Daniel Mowrey, Ph.D. reviews studies using Ginkgo. In one study patients were given 120 mg of ginkgo, daily, for twelve weeks. Patients reported a definite improvement in alertness and memory. Ginkgo and glutamine provide an effective combination for those with problems in concentration, memory, and staying on task. Ginkgo promotes an increased nerve transmission rate, and improves synthesis and turnover of brain neurotransmitters.

For those with alcohol cravings, Dr. Roger Williams pioneering in glutamine research found 3,000 to 4,000 milligrams of glutamine, daily, will stop the craving for alcohol and decrease the craving for sweets. Since pure pharmaceutical glutamine, such as Super Glutamine, is tasteless, and mixes readily with water or any cool liquid, patients find it easy to take. Patients also reported a lift from fatigue, both mental and physical. One alcoholic stopped drinking when glutamine was administered daily. Two years later the patient was still free from the craving for alcohol. He maintained a nutritional support program. Dr. Lorene

Rogers, researcher at the University of Texas, Clayton Foundation, reported several cases in which glutamine was successful and placebos ineffective. Glutamine was given to one group of alcoholics and placebos to the other. The group taking at least 3,000 milligrams of glutamine daily were free of alcohol craving.

The brain converts glutamine to energy, and with glutamine the brain's main fuel, it converts the glutamine to GABA, with the help of magnesium. Without continued high energy in the brain, the rest of the mind and body will NOT function properly. The brain requires a huge supply of glucose and oxygen in order to perform properly. This energy supply transports via the bloodstream. Proper circulation ensures the brain has the glutamine (energy) it needs.

Unfortunately, foods are not a good source of glutamine. The foods highest in glutamine include meat, chicken, and eggs, but in the RAW form. Cooking or heating inactivates glutamine, so your best source is supplement form.

Glutamine is available in capsule and powder forms. If using powder form, put in *cool* water or juice. Heat destroys glutamine.

Glycine

Glycine is a nonessential amino acid, and has the simplest structure of all the amino acids resembling glucose (blood sugar) and glycogen (excess sugar converted in the liver for storage). Glycine is sweet to taste, and can be used as a sweetener. It can mask bitterness and saltiness. Pure glycine dissolves readily in water. As the third major inhibitory neurotransmitter in the brain glycine readily passes the blood-brain barrier. The body needs glycine for the formation of DNA, collagen, phospholipids, and for the release of energy.

According to Ronald Kotulak in his book, *Inside the Brain,* glycine "helps trigger brain cells to fire electric charges and speed learning." Glycine helps spasticity and seizures, and is involved in behaviors related to convulsions and retinal function. If is taken orally, glycine increases the urinary excretion of uric acid, and is possibly a useful adjunct to gout.

Glycine is an essential intermediate in the metabolism of protein, peptides, and bile salts. Liver detoxification compounds, such as glutathione, must have glycine present for formation. Glycine removes heavy metals such as lead from the body, and also decreases the craving for sugar. In many cases, replacing sugar on foods such as cereal. Glycine has been shown to calm aggression in both, children and adults. When combined with GABA and glutamine, glycine influences brain function by slowing down anxiety-related messages from the limbic system. Glycine is effective in alcohol withdrawal as it decreasing the craving for sugar.

As a very nontoxic amino acid, glycine can be used by both children and adults. Glycine is found in high concentrations in meats and wheat germ.

Usual dosage range is 500 to 3,000 mg, per day, in divided doses.

Histidine

Histidine, one of the essential amino acid, is required in large amounts in infants. Histidine is necessary for the maintenance of myelin sheaths of nerves, has vasodilating and mild anti-inflammatory properties. The neurotransmitter, histamine, derives from histidine. Histidine promotes large increases in brain histamine content, especially in the hypothalamus.

People with abnormally high amounts of histamine often demonstrate a history of psychiatric problems ranging from mild to severe. People with chronic pain and fibromyalgia demonstrate high histamine levels represented by joint swelling. High histidine and histamine levels are often seen in patients with obsessive-compulsive disorders, depression, and phobias. Low blood-histamine levels are found with rheumatoid arthritis and Parkinson's disease.

Best food sources for histidine include pork, wheat germ, sausage, chicken, turkey, duck, ricotta cheese, and cottage cheese.

Lysine

Lysine is an essential amino acid, and must be obtained by the diet, as it cannot be produced by the human body. In the body, lysine is a critical protein required for growth, tissue repair, and production of hormones, enzymes, and antibodies. Additionally, it helps reduce the incidence of herpes outbreaks.

Symptoms of lysine deficiency include fatigue, inability to concentrate, irritability, bloodshot eyes, retarded growth, anemia, hair loss, and reproductive problems.

Lysine is effective against herpes because it reduces viral growth. It suppresses the virus by restoring the balance of nutrients retarding viral growth. The proper balance of lysine to arginine ratio helps suppress the virus. (See section on herpes.)

Good food sources of lysine include eggs, meat, fish, milk, cheese, and yeast. Cereals, rice, millet, wheat, and sesame seeds contain very little lysine. Amounts required for optimum health varies widely from person to person from 500 to 1,600 mg per day, depending on their particular biochemistry. If outbreaks of herpes occur, increase the amount to 3,000 mg, daily, until outbreak subsides.

Methionine

Methionine, an essential amino acid, represents one of the sulfur containing amino acids. Methionine is a methyl donor, critical for the formation of many important substances such as nucleic acids, epinephrine, choline, lecithin, carnitine, melatonin, collagen, serine, creatine, and deanol. Additionally, methionine can be a detoxifying agent assisting the removal from the body of heavy metals such as lead. Methionine is necessary for selenium to be absorbed and utilized in the body. As an antioxidant, it helps protect the body from effects of radiation. Normal metabolism of homocysteine require B6 and methionine. Excess homocysteine can cause plaque formation in the arteries, leading to cardiovascular disease. If supplementing with methionine, *always add B6 and folic acid* to prevent a buildup of homocysteine. Methionine can be synthesized into cysteine, cystine, and taurine, if sulfur is present.

Excess methionine has been suggested in one type of schizophrenia, while low levels are seen with depression. Supplementing with methionine helps lower histamine levels in the body, and sufferers of allergies, asthma, and chronic pain may find methionine supplementation helpful. Heroin addicts often have low pain thresholds and high histamine levels. Methionine helps lower the excess histamine levels usually present during heroin, amphetamine, or barbiturate withdrawal. In some depressed patients, methionine lifts depression with supplementation of 1 gram of methionine, morning and evening. Compared to MAO-inhibitor antidepressants, methionine proves more effective.

Good food sources for methionine include sunflowers, pork, sausage, duck, wild game, lentils, pumpkin and sesame seeds, avocado, cottage cheese, cheese, and wheat germ.

Phenylalanine

Phenylalanine, an essential amino acid, functions as the parent substance, or precursor of tyrosine. Phenylalanine converts to tyrosine in the liver.

Phenylalanine Pathway

Phenylalanine → Tyrosine → Dopamine → Norepinephrine → Epinephrine

Phenoketonurics (PKU) cannot convert phenylalanine into tyrosine because PKUs lack the enzyme, phenylalanine hydroxylase.

The formation of the hormone, thyroid, requires phenylalanine. Although phenylalanine is not found in the brain, it resides many brain peptides, proteins, and neurotransmitters. Phenylalanine is the raw substance that produces several compounds of the catecholamine family responsible for the transmission of nerve impulses, assuming an adequate supply of phenylalanine. Norepinephrine, a major neurotransmitter, derives from tyrosine or phenylalanine. The amount of norepinephrine available to the brain is predisposed by the amount of phenylalanine or tyrosine available. Phenylalanine is one of the few amino acids readily converted into brain compounds like norepinephrine that control a person's mood. Phenylalanine or tyrosine, helps give a positive, uplifting effect on mood, alertness, and ambition. Often, this amino acid is deficient in depressed people. Phenylalanine can also stimulate the release of CCK, cholecystokinin, that in effect, turns off the appetite.

Other phenylalanine derivatives such as epinephrine are excreted at the nerve terminals in the hypothalamus, and norepinephrine is excreted at the sympathetic nerve endings, giving rise to the fight-or-flight response. Norepinephrine is stored in presynaptic vesicles in certain central synapses. During times of stress, the body's adrenal glands are under immense pressure to produce epinephrine and norepinephrine. Often, they become low or depleted. This depletion can lead to depression and stress which can cause pain, anxiety, uncertainty, and fear.

Supplementing with phenylalanine or tyrosine helps to increase the level of norepinephrine in the brain. Many antidepressants work by increasing or manipulating the norepinephrine level in the brain. Often the drugs work by blocking the norepinephrine from re-entering the vesicles or pouches found at the synapse. The natural way to normalize the brain levels of norepinephrine is with supplementation of tyrosine or phenylalanine. Therapeutic dosage ranges from 500 to 1,500 mg per day. Phenylalanine should be used with extreme caution in hypertensive patients; and always take with food. People taking MAO inhibitors or tricyclic antidepressants should not use phenylalanine or tyrosine.

The DL-form of phenylalanine, or DLPA, was found to be effective in the treatment of pain, and for the depression resulting from the pain. DLPA increases the production of PEA, norepinephrine, and endorphins. PEA is a neurotransmitter-type substance with structural resemblance to amphetamine, a stimulant drug. Endorphins are the morphine-like neurotransmitters that decrease pain and gives a sense of well-being. DLPA increases endorphins by preventing the breakdown of the endorphins in the brain, so they remain there longer. If you use DLPA, the suggested amount is 750 mg, four times daily.

Food sources of phenylalanine include dairy products such as cottage cheese, milk, other cheeses; meats such as chicken, turkey, and duck; and pecans, sesame seeds, lima beans, and lentils.

Therapeutic dosages of DLPA range from 500 to 3,000 mg per day, divided.

Proline

As a nonessential amino acid, proline is required for the formation of collagen; but Vitamin C must be present. The body can manufacture proline from ornithine or glutamic acid and, if needed, the body convert back into orthinine.

Elevation of proline may be found in alcoholics with cirrhosis, and in some patients with depression or seizure disorders. Convulsions, elevated blood calcium levels, and osteoporosis may be caused by excess proline from a genetic error.

Good food sources of proline include cottage and ricotta cheeses, eggs, pork, luncheon meats, wheat germ, turkey, and duck.

Supplemental dose range ranges from 500 to 1,000 mg, with Vitamin C.

Serine

Serine, a nonessential amino acid, is synthesized from glycine with the presence of folic acid and B6. Serine is involved in DNA synthesis. Serine, in combination with carbohydrates, may form glycoproteins. As an immunosuppressive, serine may possibly be helpful in auto-immune diseases. Serine is required for the formation of choline, ethanolamine, phospholipids and sarcosine necessary for the formation of neurotransmitters, and to stabilize cell membranes. Phospholipids are made from phosphatidylserine, and requires the presence of methionine and folic acid. Excess serine may cause psychosis and elevation of blood pressure.

Phosphatidylserine (PS) is a component of brain cell membranes. Studies support phosphatidylserine's importance in brain functions such as memory and alertness, as well as enhanced function of the aging brain. Often, aging involves alteration of structure and biochemical changes within the brain. These can include changes in neuronal membrane lipid makeup and enzyme activity, reduced production and metabolism of neurotransmitters, and loss of nerve synaptic connections. In one study, 35 patients (19 males and 16 females) with prevalent involvement of mental functions associated with behavioral changes, were treated with PS (300 mg, daily) for a period of two months. The therapeutic activity of PS was evaluated through neuropsychological tests and behavioral rating scales. Results demonstrated PS was a beneficial treatment for mild to moderate deterioration of cognitive function. Another study involved probable Alzheimer's disease patients. The study confirmed that PS treated patients improved several cognitive functions; and suggested early stage Alzheimer patients may benefit most from PS supplementation with a dosage of 100 mg, three times, daily.

Good food sources include cottage and ricotta cheeses, wheat, wheat germ, pork, luncheon meat, turkey, sausage, peanuts, and soy.

Taurine

Taurine is now classified as a conditionally essential amino acid in the adult; but in infants and children, taurine is an essential amino acid. Taurine is one of the sulfur amino acids, and in the adult is synthesized from cysteine and methionine, provided B6 and some zinc are present. Taurine is found throughout the body abundantly in the heart muscle, olfactory bulb, central nervous system, and brain—hippocampus and pineal gland. Taurine participates in a multitude of functions in the body involving the gallbladder, brain, heart, eyes, and vascular systems.

In the heart, taurine is the most concentrated amino acid. It is involved in the heart muscle contractility and rhythm. After a heart attack the levels of taurine often decrease dramatically. In some cases the levels drop to one-third the normal. In one study, congestive heart failure (CHF) patients were given 4 grams of taurine per day for a month. Improvement was seen in 19 of the 24 patients. CHF patients frequently show increased levels of taurine because the body tries to naturally correct the problem metabolically. Taurine seems to assist in CHF by acting as a diuretic and a heart stimulator with doses of 2 grams per day. Taurine and magnesium levels drop dramatically whenever heart arrhythmias occur, and replenishing both assists in controlling heart arrhythmias. Additionally, taurine helps balance the calcium and potassium levels in the heart.

In the retina of the eye taurine is present in high concentrations. Retinitis pigmentosia patients show decreased levels of taurine. However, no documented cases of blindness have been attributed to taurine deficiency.

Normal brain development in infants requires taurine. It protects and stabilizes the brain's fragile cell membranes. Taurine also acts as an inhibitory neurotransmitter in the brain. It closely resembles the structure and metabolism of the other inhibitory neurotransmitters, GABA and glycine. Taurine proves effective in the treatment of epilepsy, acting as an anticonvulsant. The levels are lower than normal in over half of the amino acids, if epilepsy is present, but the levels of taurine are higher than normal, except in the cerebrospinal fluid. Usual dosage of taurine for epilepsy is 3,000 mg, per day, with a non-protein meal. Taurine helps people with tics or other spastic conditions.

A deficiency of taurine has been demonstrated in some patients with depression. A deficiency can add to chemical sensitivity and decrease the body's ability to detoxify chemicals.

Taurine is necessary for the formation of one of the bile acids and for proper functioning of the gallbladder. The bile may be a route of excretion of chemicals detoxified by the body. Taurine is sometimes called upon to help control inflammation or infection.

Taurine, or a modified taurine, may someday supersede synthetic tranquilizers.

Best food sources of taurine include meats, especially organ meats, and fish. The need for taurine increases whenever you experience more stress than usual or have an illness. Supplementation is necessary as the taurine need becomes greater than what can be obtained from the diet, alone.

Usual dosage of taurine is 500 to 3,000 mg per day, preferably on an empty stomach. Women require more taurine than men since men have higher enzyme levels.

Threonine

Threonine is an essential amino acid, and is the precursor to brain glycine. It is required for proper digestion and intestinal tract function. Threonine breaks down into glucose, and into the amino acids glycine and serine.

Deficiency of threonine suppresses the immune system. It has been helpful in multiple sclerosis cases and in some patients with agitated depression and mania.

Good food sources includes pork, turkey, wheat germ, and cottage and ricotta cheeses.

Tryptophan

Tryptophan is an essential amino acid, and must be obtained in the diet. It ultimately breaks down into serotonin, the calming neurotransmitter in the brain. Serotonin helps us feel calm, relaxed, and in control.

Tryptophan Pathway

Tryptophan → 5-HTP → Serotonin

Tryptophan is a precursor of serotonin. Serotonin is synthesized from tryptophan. Serotonin is a brain neurotransmitter, platelet-clotting factor, and neurohormone found in the organs throughout the body. Tryptophan is essential to maintaining the body's protein balance. When food that is protein deficient or lacking tryptophan is fed to growing or mature individuals, such foods fail to replace worn-out materials lost by the body during the organic activities of its cells, tissues, and organs. The amino acid tryptophan is exhausted by the vital activities of the body and, in turn, must be replaced to prevent atrophy of the body's structures.

One of the few substances capable of passing the blood-brain barrier, tryptophan plays a variety of important roles in mental activity. When tryptophan intake is deficient, especially during periods of stress, serotonin levels drop, causing depression, anxiety, insecurity, hyperactivity, insomnia, and pain. The body requires ample supplies of Vitamin B6 for the formation of tryptophan.

Tryptophan's role in behavior has been demonstrated by the number of mental functions it directly influences. Serotonin produces a relaxed, calm, secure, mellow, and morphine-like analgesic feeling. Only 1 to 2% of all serotonin in the body is found in the brain. Hyperactive children/adults have a low serotonin level. Aggression reflects one of the most widely recognized signs of reduced serotonin. Supplements containing tryptophan and Vitamin B6 can correct some of the biochemical disorders related to aggression.

Another significant finding in studies done with tryptophan demonstrated that low levels of serotonin could play a part in the development of depression. Combining tryptophan (1,000 mg) with tyrosine in doses of 3,000 mg per day, at bedtime, can mimic the effects of most antidepressants. Tryptophan is useful in unipolar depression or constant, low-grade depression with no highs or lows.

Both depression and pain can have profound effects on a person's ability to fall asleep. Difficulty in falling asleep can be caused by low serotonin levels. But tryptophan has been shown to effectively solve insomnia problems, reducing the time needed to fall asleep, and increasing the number of hours spent sleeping. The usual dosage is 500 to 1,000 mg, taken one hour prior to bedtime with a carbohydrate such as orange juice or fruit.

Because serotonin is a neurotransmitter, it is one of the most important chemicals to help control moods. Best of all, tryptophan is safe, and is a natural relaxant and tranquilizer of the central nervous system. The body has no difficulty in rapidly metabolizing and clearing it from the body. As an essential amino acid necessary for life, tryptophan is the sole precursor for serotonin. It does not simply pass through the gut into the brain, to become serotonin. It must compete with five other amino acids—tyrosine, phenylalanine, leucine, isoleucine, and valine—at the blood-brain barrier. In order to increase the amount of brain serotonin, the ratio of tryptophan must be elevated out of proportion to the competing amino acids. Metabolism and protein intake may alter this ratio.

About 90% of serum tryptophan is bound to albumin. Free fatty acids (serum) share the same albumin binding sites. Changing both the blood-sugar level and insulin may increase and decrease the proportion of free tryptophan that has access to the brain. In the total serum amino acid profile, the ratio of tryptophan to the nutrient amino acids has been elevated in each instance. Tyrosine has also been elevated in each instance. About 1% of the ingested tryptophan is metabolized to serotonin. About 90% of the tryptophan is metabolizes through kynurenic acid to nicotinic acid.

The neurotransmitters directly depend on dietary tryptophan and other amino acids. Circadian rhythms effect the amino acid utilization in the nervous system. Circadian rhythm is a specific type of periodicity for the uptake and utilization of substances. This has recently been shown with use of tryptophan in the treatment of insomnia. When tryptophan is used during the day, it does not seem to induce sleep, only a calm, relaxed state; but when taken near bedtime, it seems to induce sleep, as shown in sleep studies done at several medical centers. This seems to indicate that tryptophan's uptake across the blood-brain barrier corresponds to the circadian rhythms of the sleep cycle. Its absorption and conversion in the brain to serotonin more effectively occurs during times when a person would normally sleep. This is why it is clinically suggested to administer tryptophan at bedtime, if used for treating sleep disorders. Conversely, tryptophan or tyrosine should be used during the day to treat certain forms of depression. Liquid homepathic serotonin used 3 to 4 times daily will elevate the serotonin level. Melatonin also elevates the serotonin level, and is effective for sleep problems. Melatonin is produced by the pineal gland in the brain, and is a neurohormone.

Two researchers in England compared the antidepressant effects of tryptophan and Tofranil. (Tofranil, or imipramine, is a drug commonly used for depression.) Both groups of patients with depression improved. The study revealed that tryptophan was just as effective as the laboratory-produced drug, and there were no side effects from the tryptophan. Conversely, the side effects for the Tofranil group included blurring of vision, dryness of the mouth, low blood pressure, urinary retention, heart palpitations, hepatitis, and seizures.

Tryptophan is obtained in the diet every day. Many rich natural forms of tryptophan include: bananas, green leafy vegetables, red meat, pork, turkey, dairy products, pineapple, avocados, eggs, soy, sesame and pumpkin seeds, and lentils. Large doses of tryptophan, when combined with niacinamide and Vitamin B6, can enhance the conversion of tryptophan to serotonin.

Currently, tryptophan is available only by prescription. It was removed for sale in 1989 because of a contaminated batch that caused EMS (eoinophilia-myalgia syndrome). The F.D.A. determined the cause was tryptophan and not the contaminated batch. It reclassified tryptophan as an unapproved experimental drug, and ordered recall of all products except where tryptophan occurred naturally. To date, tryptophan is still banned for sale in the U.S.

But within the last year, 5-HTP, or 5-hydroxytryptophan, has become available. 5-HTP derives from griffonia seeds, a member of the legume or bean family. 5-HTP, about 10 times stronger than tryptophan, is one step closer to serotonin. Suggested dosage is 50 to 300 mg of 5-HTP, daily.

One study compared 5-HTP to Luvox, an antidepressant. Subjects with depression were given 100 mg of 5-HTP, three times daily, or 150 mg, of Luvox, three times daily. Evaluations were done at 2, 4, and 6 weeks. After 2 weeks, both groups reported a significant reduction in depression. By week 4, 15 of 36 5-HTP patients and 18 of 33 Luvox patients reported at least a 50% improvement in depression symptoms. Final assessment demonstrated the 5-HTP patients had the greatest improvement and the least amount of treatment failures. Another study involving endogenous depression (arising from within the individual, in all likelihood genetic) demonstrated marked improvement or cure in 69% of patients receiving 5-HTP.

Tyrosine

Tyrosine is the first breakdown product of phenylalanine, and is considered a nonessential amino acid because the body can make it from phenylalanine.

Dr. Gelenberg, at Harvard Medical School, determined tyrosine more effective than antidepressants for relief of depression. To rapidly increase the norepinephrine level, use tyrosine. Because it is one step closer to norepinephrine, you feel the effect more rapidly. Suggested amount of tyrosine is one 850 mg capsule of pharmaceutical grade tyrosine, three times daily. *Do not take tyrosine with MAO inhibitors or tricyclic antidepressants.*

See the phenylalanine section for metabolism and specifics.

B.N.C.

Balanced Neurotransmitter Complex

The brain communicates through neurotransmitters, the chemical language of the brain. A balanced neurotransmitter complex contains the amino acids in a special blend that nourishes the brain. An unbalanced diet, anxiety, chronic pain, depression, and grief, plus other factors, can contribute to disturbances in amino acid metabolism. B.N.C. contains the special amino acid mix of phenylalanine, leucine, valine, histadine, arginine, lysine, isoleucine, alanine, glutamine, methionine, threonine, alpha-ketoglutaric acid, pyridoxal 5'phosphate (B6), and chromium picolinate. These 12 amino acids, plus activity agents, can be taken on a daily basis without fear of creating an amino acid imbalance. B.N.C. can be combined with other amino acids for a total orthomolecular approach. Research has shown substantial improvement in chronic fatigue patients using the B.N.C. complex, CoEnzyme Q10, and Mag Link. B.N.C. should be taken on a daily basis to correct impairments in biochemistry that can either cause or complicate health conditions. Amino acids, because of their intimate involvement in metabolic regulation, proves very useful therapeutic agents that reverse biochemical impairments related to amino acid metabolism. B.N.C. has been instrumental in correcting mental and stress-related disturbances, food and chemical intolerances, learning disabilities, frequent headaches, fibromyalgia, chronic fatigue, mental and emotional disturbances, hyperactivity, and some neurological disorders. B.N.C. is available in capsule and powder forms.

Amino Acid Cofactors

Timed Released B6

Rodex B6

Supplementation with timed-released B6 disperses B6 over a period of 8 to 9 hours. This timed-release B6 helps prevent neurotoxicity which can occur in doses greater than 500 mg over a prolonged period of time.

P5'P or Pyridoxal 5'Phosphate

(Active B6)

Pyridoxal 5' Phosphate, or P5'P, the biological (active) form of B6—is necessary for the utilization of all amino acids, proteins, fats, and carbohydrates. If P5'P is not present, increased excretion of most amino acids occurs as well as increased formation of abnormal amino acid metabolites.

Unlike B6 there is no fear of toxicity with P5'P. Reaction to MSG may indicates a deficiency of Vitamin B6. Reactions to MSG may effectively be prevented with supplementation of B6.

Magnesium

Magnesium is an essential cofactor in over 300 enzyme reactions in the body. Many Americans are deficient in magnesium, and do not get enough from their diets. Magnesium and B6, or P5'P, must be present or the body cannot assimilate and properly use amino acids.

Amino Acids in Therapy

Addiction (Alcohol)

Alcoholism is a disease of chemical dependency. It is addictive, abusive, and eventually becomes destructive. Alcoholism ranks with stress, mental illness, and heart disease as one of the major problems in the U.S. Alcoholism develops from a combination of factors—psychological, physiologic, genetic, and environmental. Presently, records show alcohol is the most abused drug in the U.S., and the problem is on the increase, especially among teenagers.

Researchers have established multiple nutrient deficiencies in those craving alcohol. Many are predisposed to alcoholism because of genetics. Those who have alcoholic parents or grandparents will have the same brain deficiencies which can lead to addictive behavior.

Dr. Roger Williams and his colleagues at the Clayton Foundation for Research at the University of Texas, established the vital research concerning the amino acid L-glutamine. Those craving alcohol have a definite glutamine deficiency. Dr. Williams and his associates observed that glutamine protects individuals against the poisonous effects of alcohol, and that it stopped the craving for alcohol. They studied all the properties of glutamine and GABA, and found those who had addictive behaviors had deficiencies.

Glutamine and GABA will decrease the craving for alcohol. Pure glutamine is tasteless, and can be mixed with food, water, or taken in capsule form. One alcoholic who was part of the study stopped drinking after he was given 3,000 mg of glutamine, daily, along with other necessary nutrients. Several studies demonstrated glutamine effectively reduces the craving for sweets. The same appetite center in the brain and hypothalamus protects against alcohol craving. Glutamine is the third most abundant amino acid in the blood and brain. Glutamine provides a major alternative fuel source when blood-sugar levels are low. Amino acids create the needed neurotransmitters to enhance the brain chemistry. A strong nutritional program is of utmost importance for the control of alcoholism and addictive behaviors.

Suggested Nutritional Support

Glutamine powder or caps - 1,000 to 4,000 mg per day, divided.

Rodex B6 (timed release) - 1 (150 mg) capsule per day.

B Complex - 1 capsule per day.

Mag Link - 4 to 6 per day, divided. Take to bowel tolerance then decrease by 1.

Ester C - 2,000 to 3,000 mg daily, divided.

Tyrosine 850 - 1 (850 mg) twice per day.

Anxiety Control 24 - 2 twice to three times per day, as needed, for stress and anxiety.

T-L Vite (multi-vitamin) - 1 capsule with breakfast.

DLPA 750 mg - two, twice daily. (Do not use if you are taking MAO inhibitor or tricyclic antidepressants).

Cal, Mag, Zinc - 4 per day, at bedtime.

Liquid Serotonin - 10 to 12 drops, four times per day.

5-HTP - 1 to 2 capsule (50 mg) an hour before bedtime, with juice. Do not take if you are taking an SSRI antidepressant.

Word of Warning . . .

"Drugs such as major tranquilizers should be considered as temporary crutches, to be used until the biochemical imbalances are slowly corrected by nutrient therapy. Anti-psychotic drugs, if continued at high doses for many months, may produce tardive dyskinesia—a delayed impairment of voluntary motion causing incomplete or partial movement. Manganese taken daily is helpful, as is a daily dose of deanol, which builds up acetylcholine, then normal working hormone in muscle contraction."

Excerpted from *Nutrition and Mental Illness,* by Carl C. Pfeiffer, Ph.D., M.D.

Adult A.D.D. / Hyperactivity

Recently, there has been an increase in the number of adults who feel they have A.D.D. or A.D.H.D., major problems they describe include their inability to concentrate, complete tasks, or stay focused. Many of them have turned to the powerful drug, Ritalin. Stimulant drugs such as Ritalin impair brain function and have no beneficial effect on the brain. According to Peter Breggin, M.D., in his book, *Talking Back to Ritalin,* Ritalin can either shrink or limit the growth of some areas of the brain, just as many psychiatric drugs can cause brain dysfunction and damage.

There are 16 adverse reactions listed in the PDR (Physicians Desk Reference) for Ritalin. The most common adverse reactions are nervousness and insomnia. Other reactions include skin rashes, fever, anorexia, dizziness, palpitations, headaches, dyskinesia, drowsiness, blood pressure and pulse changes, pulse >100, angina, cardiac arrhythmias (heart rhythm disturbances), abdominal pain, and weight loss. Given this information, there should be no question about your decision.

There is no such thing as a Ritalin deficiency! There are many reasons why as we age our memory might not be what we feel it should, but Ritalin is *not* the answer. Your brain cannot be deficient of Ritalin. If you cannot concentrate or stay focused, you need neurotransmitters. The following nutritional support program will help you nourish your brain so you feel better and think more clearly.

Suggested Nutritional Support
Brain Link Complex - 1 scoop in the morning, depending on weight; if over 200 pounds, 2 scoops.
Glutamine - 500 mg capsule, twice daily.
Mood Sync** - 1 capsule in the morning, afternoon, and evening.
Mag Link - 1 tablet, three times per day.
Ginkgo Biloba - 40 mg capsule, twice daily.
5-HTP** - 1 capsule, 30 minutes prior to bedtime.

Special Note: If your symptoms become acute after you eat certain foods, your problems could be related to food allergies. Millions of children and adults are diagnosed with A.D.D. or A.D.H.D. when their problem is really food allergy related. For detailed information regarding A.D.D. and A.D.H.D., read my book, *Control Hyperactivity/A.D.D. Naturally.* It is available at your health food store or through the Pain & Stress Center at 1-800-669-2256.

** Do not take if you are taking an SSRI antidepressant.

Allergies / Sinus Problems

Allergies are the cause of sinus congestion and problems in about 90% of the cases. If you experience constant congestion and pressure, allergies could be the problem. Explore the possibility of chemical antigens as well as airborne. Food allergies can also cause congestion, stuffiness, headache, palpitations, upset stomach, fatigue, etc.

Suggested Nutritional Support

NAC - 1 600 mg capsule, twice to three times per day.

P5'P - 100 mg, morning and evening.

Tyrosine (500 mg) - 1 in morning and 1 in evening. Do not take if you are taking a MAO inhibitor, tricyclic antidepressant, or if you have had a malignant melanoma.

Vitamin C (Ester C) - 2,000 mg, three times per day.

Mag Link - 4 to 6 tablets per day, divided. Magnesium helps prevent release of histamine. If diarrhea or loose stools occur, take tablets more divided or decrease dosage by 1.

Pycnogenol - 300 mg per day. Pycnogenol, or OPC, has great antioxidant value due to its free radical scavenging ability.

BHI Sinus (homeopathic) - 1 tablet every 15 minutes until relief; then 1, four times per day.

BHI Headache (homeopathic) - For sinus headache, use 1 tablet every 15 minutes until relief, then 1 four times per day.

Euphorbium Nasal Spray - Use as needed for nasal congestion. Euphorbium helps normalize nasal tissues, and does not have a rebound effect.

Oil of Oregano - Use 6 number of drops in fruit juice for chronic congestion. Make sure the oregano is ***wild oregano***. In addition, use Oreganomax; 2 capsules, three times per day, for chronic allergies and sinusitis.

Carpal Tunnel Syndrome

Carpal tunnel syndrome (CTS) occurs due to an entrapment of the median nerve as it crosses the wrist. CTS causes numbness and tingling in the thumb and middle three fingers, and can lead to pins-and- needle sensations in the entire hand. Symptoms generally worsen at night and early morning.

Repetitive stress/strain injuries (RSI) have now become known as the workplace curse of the 90s. RSI symptoms can include everything from neck and shoulder to back strain. The cause is prolonged repetitive movements involved in certain occupations or recreational activities. Manual tasks during which repeatedly flexed and extended the wrist intensify CTS with swelling and compressing the wrist nerves.

Suggested Nutritional Support

Rodex B6 - 1(150 mg) timed-release B6 capsule, daily.

DMSO - Apply topically to wrist, as needed, for swelling.

Mag Link - 2 tablets, twice daily, for muscle tension and strain.

Glucosamine - 500 mg capsule, three times daily.

Powerelief - 1 or 2 capsules every 6 hours, as needed, for pain.

Bromotol (bromelain) - 500 mg, twice daily, to support tissue repair and reduce swelling.

Ester C - 1,000 mg, morning and evening.

Magnetic wrist tube or wrap support - wear daily.

Ice Pack - Use an ice pack such as SofTouch packs, at night, to reduce swelling and pain.

Mood Sync - 1 capsule, twice daily, for mood support.

Chronic Pain

Pain is the hot poker in your lower back, the jackhammer pounding into your skull or temples, the invisible vice twisting your fingers and hands into grotesque and useless lumps, the fire burning in your neck and shoulders. Pain sends an estimated 120+ million Americans to doctors, pain clinics, and chiropractors every year. Pain is the stone cross no one has the strength to bear. It cripples our lives and drives millions to addictions to addictive prescription drugs and alcohol.

Every year, the American public spends 6 billion dollars for analgesics (painkillers) and therapies of every description, but the pain is still there. Pain is one of the most serious problems not only in the U.S., but the world. Back pain afflicts 85 million, arthritis 56 million, migraines 30 million, and 900,000 live with the hell on earth of cancer pain.

There are no magic bullets, no miracle drugs, no overnight cures. If anyone tells you they have a cure for pain, run, don't walk away. If a physician tells you surgery is the answer, get a second opinion and then make sure they review all of the potential post surgical problems. If you are offered an array of medication, consider the long term effects because drugs only treat the symptoms. Your goal is to treat the source, the master controller, that sends signals to every corner of the body, your brain!

Remember, whatever the brain tells the body to do, it does! Pain research at John Hopkins University indicates a great deal of chronic pain and depression reflects the patient's inability to produce enough of certain brain chemicals. This research points to new methods of treating pain and depression, without using harmful drugs.

Studies at other major universities have shown the brain does produce many hormone-like chemicals that have a close functional resemblance to morphine. These morphine-like chemicals are called endorphins (endogenous morphine) because they are produced by the body (endogenous), and are similar to morphine.

Endorphins regulate pain and control the transmission of pain signals. Endorphins are inhibitory neurotransmitters in the brain and nervous system. They slow down the transmission of pain information from the limbic system to the cortex. Endorphins and other neurotransmitters help link together the one hundred plus billion neurons of the human brain into an incredibly complex network. From this complex network comes our natural pain killers, endorphins. Endorphins have been shown to be more powerful than morphine. Are they the answer to pain control? Yes and no.

Endorphins are not drugs, but a simple nutritional amino acid, phenylalanine. The pain studies involved DL-Phenylalanine, or DLPA. DLPA is not a drug. It does not actually block pain, itself. DLPA works by protecting your own naturally produced endorphins, effectively extending their life spans in the nervous system. By extending their lifespans, pain relief results. DLPA is a natural healer. It helps the body heal itself. Depression loves pain, pain loves depression; but DLPA has the ability to elevate the brain chemistry so that you do not feel depressed and in more pain. DLPA can do this because it is an inhibitory neurotransmitter.

DLPA is now being used clinically throughout the U.S. in major pain clinics, for pain and depression. Many of the millions of people suffering from pain do so because of stress. Stress energizes pain, pain fuels stress. We see patients every day for stress-induced pain taking a combination of pain medications and antidepressants. *Harrison's Principles of Internal Medicine* states 50 to 80% of all pain is stress induced. This means there is no pathology; the pain is there, but there is no disease.

Suggested Nutritional Support:

DLPA 750* - Take 5 minutes after meals, 3 times per day. Keep in mind DLPA takes from 2 days to 3 weeks to take effect. If your doctor has prescribed medication, you do not have to stop taking it. DLPA can greatly enhance the effectiveness of aspirins and analgesics.

Bromotol (bromelain) - 2, 500 mg tablets morning, and 2 evening.

Rodex B6 (time released) - 1 capsule, 150 mg before breakfast.

Mag Link - 2 tablets mid-morning and 2 mid-afternoon. If no diarrhea or loose stools occur, try increasing to a total of six per day. Take to bowel tolerance, then decrease by 1.

Powerelief* - 2 caps, twice to three times per day.

Boswella - 1 or 2 capsule(s), mid-morning and mid-afternoon.

Anxiety Control 24 - 2 capsule in morning, and 2 in evening.

TL Vite - 1 capsule with breakfast

Mobigesic - 1 tablet every 4 to 6 hours.

Ester C - 1,000 mg, four times per day.

* DLPA should not be taken by pregnant or lactating women, those with PKU, if you are taking MAO inhibitors or tricyclic antidepressants, or if you have had a malignant melanoma.

5-HTP - Start with 1 (50 mg) 30 minutes before bedtime; increase to 2, if needed. Do not take if you are taking an SSRI antidepressant.
OR use
Melatonin - 1 (3 mg) capsule an hour before bed, for sleep.

• Try to use a relaxation tape at least 30 minutes every day.

Depression

There is hope for those suffering from chronic depression, the lonely emotion. Depression is very treatable without toxic antidepressant drugs.

Major Symptoms of Depression:

- Passive negativity
- Oversleeping
- Constant indigestion
- Dry mouth
- Compulsive eating, especially carbohydrates
- Appetite loss
- Constipation or diarrhea
- Inability to make decisions
- Loss of confidence or self esteem

Depression can coexist with other disorders both physical and psychological. Each disorder can feed off the other such as chronic pain, fibromyalgia, or headaches.

If you have a chemical imbalance causing your depression, you have a deficiency of norepinephrine or serotonin. These two neurotransmitters are the major neurotransmitters that control mood in the brain. Drugs work by manipulating these neurotransmitters—usually by increasing the amounts of norepinephrine or serotonin.

Tyrosine, because of its role in assisting the body to cope physiologically with stress and building the body's natural store of adrenaline, deserves to be called the stress amino acid. Stress exhaustion needs tyrosine which is converted to dopamine, norepinephrine, and epinephrine. The use of tyrosine in depression increases levels of serotonin and neurotransmitters. These help restore a sense of well-being.

Tyrosine was first used in psychiatry for medication-resistant depression. Dr. A. J. Gelenberg (1980), of the Department of Psychiatry at Harvard Medical School, used tyrosine to treat patients who presented with depression. These patients noted significant improvement with tyrosine.

5-HTP, or 5-hydroxytryptophan, boosts the serotonin levels in the brain and creates neurotransmitters that produce an inhibitory effect on the nervous system. 5-HTP is converted into serotonin in the brain. Serotonin soothes, calms, and gives you a warm feeling of contentment. If you have a deficiency of serotonin, symptoms of depression, obsessive-compulsive disorder, anxiety, pain, and migraines will be demonstrated. Both Mood Sync and Teen Link contain 5-HTP.

St. John's Wort (SJW) is known as the depression herb, and it has been used for centuries. SJW's effectiveness has been confirmed by multiple double-blind studies comparing SJW to tricyclic antidepressants such as Elavil and Tofranil. The effective dose is 300 mg, three times, a day. SJW inhibits the breakdown of neurotransmitters like serotonin and weakens inhibition of the enzyme MAO. SJW should not be used with phenylalanine or tyrosine.

People with allergies or high histamine levels will have chronic low-grade depression. Additionally for allergy sufferers, add 500 to 1,000 mg of methionine daily, divided doses, and BCAAs (Branch Chain Amino Acids), 500 to 1,000 mg, daily. These amino acids will help you feel better.

Suggested Nutritional Support

Tyrosine 850* - For chronic depression, use 850 mg, twice daily.

B6 - 100 to 150 mg daily, preferably in timed-release form, such as Rodex B6.

GABA 750 - Use 1/2 capsule three times daily, divided and dissolved in water. GABA levels of those with depression are usually low. This information surfaced in a study done by F. Petty, M.D., of the Department of Psychiatry, Veterans Medical Center in Dallas.

Liquid Serotonin - 10 to 15 drops, 3 times daily, or as needed.

Mood Sync - 1 or 2 capsules, twice to three times daily. Do not use if you are taking an SSRI antidepressant.

* Tyrosine, Phenylalanine, or DLPA should not be taken by pregnant or lactating women, those with PKU, if you are taking MAO inhibitors or tricyclic antidepressants, or if you have had a malignant melanoma.

Children
For children *under 100 pounds*, use 500 mg tyrosine,* once daily with 50 mg B6. If *over 100 pounds,* use 500 mg tyrosine, twice daily with 50 mg B6.

Teenagers
For teenagers, use 1 or 2 capsules Teen Link, twice daily. Always start with 1 capsule twice daily, and see if that lifts the depression. Only if depression is not eased, increase to 2, twice to three times daily. Do not use if you are taking an SSRI antidepressant.

*Do not take Tyrosine, Phenylalanine, or DLPA if you are pregnant or lactating, if you have PKU, if you are taking MAO inhibitors or tricyclic antidepressants, or if you have had a malignant melanoma.

Diabetes

Approximately 10 million Americans have diabetes. Diabetes is a chronic disorder of carbohydrate, fat, and protein metabolism. Diabetes occurs when the pancreas does not secrete enough insulin, or if the cells of the body become insulin resistant. As a result, blood sugar is unable to get into the cells. Diabetes can lead to multiple serious medical complications.

There are two basic types of diabetes: Type 1 and Type II. Patients with Diabetes Mellitus, or Type I, are insulin-dependent. This most often occurs in children and adolescents. Type II usually begins after the age of 40. Type II diabetics are non-insulin dependent, and comprise 90% of all diabetics.

According to Julian Whitaker, M.D. in his book *Reversing Diabetes*, some of his patients with Type I diabetes report having very stressful events in their lives that occurred six months to a year before they developed diabetes. Stress can alter the immune system, causing it to weaken. This predisposes an individual to the disease state, and often the onset of diabetes.

Suggested Nutritional Program
Always use capsule, powder, or liquid form for maximum absorption.
Anxiety Control 24™ - 1 or 2 capsules, twice per day, or as needed to decrease stress and anxiety.

Chromium Picolinate - 600 mcg daily, divided to help regulate blood sugar and decrease insulin resistance.

Gymnema Sylvestre - 300 mg, three times daily, divided before meals. Gymnema Sylvestre is a herb that aids in controlling sugar uptake and cravings. It helps to maintain normal blood-glucose levels.

Mag Link - 2 tablets twice to three times daily. If loose stools occur, decrease the dose by 1, or try spreading out the interval between doses.

Vanadium - 100 to 150 mg per day, divided. Vanadium helps with blood glucose reduction. (You may have to cut back on your oral medications. Do this under the supervision of your physician.)

Carnitine - 2 (250 mg) capsules, twice per day. Carnitine is important in fat distribution in the body. It helps reduce cholesterol and triglyceride levels to normal in the body.

Fiber - helps keep blood sugar in the normal range. Use 2 tablespoons of Fortified Flax in any food or beverage. Fortified Flax provides 4600 mg of Omega 3 (good oil) needed on a daily basis.

Rodex B6 - 1 timed-release capsule in the morning.

Taurine - 500 mg capsule, twice daily, to aid the release of insulin.

Deluxe Scavenger (antioxidant) - 3 per day, divided. Deluxe Scavenger combines CoEnzyme Q10, beta-carotene, Vitamin C, Lemon bioflavonoids, rutin, Vitamin E, Selenium, Glutathione, NAC, riboflavin, P5'P, and Vitamin B6.

Vitamin C (Ester C) - 2,000 to 3,000 mg per day, divided. Vitamin C is vital for the repair of all body tissues and scavenges of free radicals in the body.

Vitamin E - 400 to 800 I.U. daily. Vitamin E is important for the circulation, heart, neurological functions, and scavenging free radicals.

Manganese - 15 mg daily. Manganese is important for pancreas repair.

CoEnzyme Q10 - 80 to 120 mg (in capsules) per day. CoQ10 supports the immune system, helps provide increased oxygen to the heart and body, and acts as a protective factor for the heart.

Alpha-Lipoic Acid - 300 mg, twice daily; especially important for diabetic neuropathy.

Eat as many fruits and vegetables as possible. This increases fiber which helps stabilize blood sugar, and reduces the need for insulin.

Amino Acids and the Elderly

Many factors contribute to how quickly we age. The combined effects of genetic inheritance, health habits, medical history, lifestyle, sociocultural background, and environment all play a part in our aging. Many of the elderly do not take the time to cook nutritional foods. Therefore, they deplete their immune system and illness prevails.

Senior citizens are more prone to fatigue, dizziness, and falls, as their muscles do not respond as quickly as they did when they were younger. Sudden movements or exertion can increase the probability of falls.

The heart muscle begins to wear out from the stress of everyday life. The muscle becomes less elastic with age from exposure to free radicals found in air, sun, and environment. Cardiac output is reduced due to thickening and hardening of the heart valves and chambers. The overall consequences reduce oxygen delivery to the cells. In addition, plaques of fat build up in our blood vessels and arteries.

Elasticity loss in the lungs reduce vital capacity. Damage from free radicals stiffens the exchange sacs called alveoli. As a result, the air exchange is compromised, affecting the health of the body tissues.

As a person ages, their ability to digest proteins diminishes. The amount of stomach acid-and protein-digesting enzymes decreases. Up to 30% of people over 60 do not secrete any stomach acid. Since most people over 60 have a decreased ability to digest proteins, and proteins are broken down into amino acids, then supplementation with amino acids will provide a base to cover the essential amino acids needed by the body.

The digestion and elimination process slows down as we age. Poor appetite is a common complaint, due to many factors. Salivary secretion decreases by 50 to 60%. Loss of teeth, or gum disease, makes chewing more difficult and painful. Constipation may result from decreased fluid intake, lack of fiber, little or no exercise, and decreased intestinal motility.

Ingestion (%) of R.D.A. in People Over 60

	M	F		M	F
Vitamin B6	66	72	Vitamin D	51	62.5
Vitamin B12	19	31	Calcium	20	35
Folic Acid	54	72	Zinc	40	67

Probably one of the most apparent changes occurs in the skin. The skin loss or resiliency and wrinkling becomes very evident. The skin becomes drier, thinner, more fragile, and less elastic.

The bones begin to break down through bone demineralization, reduced exercise and activity, and loss of calcium from kidneys and intestines. The loss of height becomes apparent in many seniors. The mineral loss in bones makes the bones more fragile and brittle. Sometimes the bones break, causing a fall.

Function, Cognition, and Behavior Influenced by Nutrients

Nutrient	***Influence***
Taurine	Seizures
Carnitine	Cognition, depression
Phenylalanine	Catecholamines, dopamine, depression
5-HTP (Tryptophan)	Sleep, serotonin
Thiamin (B1)	Carbohydrate sensitivity
Riboflavin (B2)	Neurotransmitter control, neuropathy
Niacin (B3)	Dementia
Pyridoxine (B6)	Neurotransmitter control
Cobalamin (B12)	Dementia
Folic Acid	Dementia
Choline	Memory, Acetylcholine synthesis
Inositol	Peripheral neuropathy
Pantothenic Acid	Fatigue
Vitamin E	Parkinson's Disease
Iron	Neurophysiologic problems
Magnesium	Sleep disturbances, nervous exhaustion
Zinc	Smell, taste
Copper	Neurotransmitter control, RBC formation

Nutritional status influences all parts of the nervous system. The neurotransmitters regulate the physiologic processes, and the way the brain processes information relates to our nutritional status. In several studies done on individuals 65 and over, the results indicate cognitive tasks corresponded to the nutritional status of the individual.

Alterations in psychological and neurophysiological performances occur when a person is deprived of, or low in, the B vitamins. As an example, acetylcholine derives from the B complex vitamin, choline. Inadequate amounts of choline can affect the synthesis, release, and metabolism of acetylcholine, and alter nerve function. Serotonin, an inhibitory neurotransmitter derived from tryptophan, decreases with age. Conversely, the catecholamine family—epinephrine, norepinephrine, and the neurotransmitters derived from phenylalanine or tyrosine—decreases with aging. These changes can produce mood swings, depression, and sleep-pattern alterations. The nutritional status based on a person's unique genetic needs, determines cognitive and behavioral functioning.

In the past decade researchers have realized the relationship between the activity and function of the nervous system to the availability and metabolic activity of various nutrient-derived substances, including amino acids, vitamins, minerals, essential fatty acids, and other conditionally essential nutrients such as carnitine, taurine, and glutamine.

Suggested Nutritional Support *(Over age 60)*
For all seniors, ***always use powder, liquid or capsule form. Never take in tablet form,*** as tablets do not break down as readily, so absorption is impaired.

BNC Plus - 1 teaspoon, twice daily, in fruit juice.
B Complex Capsules - 1, twice per day.
TL Vite - 1 capsule in the morning,
(Use all three together OR use Brain Link)
***Or Use* Brain Link Complex** - 1 scoop in the morning, and 1 in evening.
Deluxe Scavenger (antioxidant) - 1, three times per day.
CoEnzyme Q10 - 50 mg capsule, once per day.
Mag Link - 2 in the morning, and 2 in the evening.
Cal, Mag, Zinc - 4 capsules at bedtime.
5-HTP - For sleep, 1 or 2 capsules an hour before bedtime, with a piece of fruit. Always start with 1 capsule, increase to 2, if needed.
***Or Use* Melatonin** - For sleep, 1 (3 mg) capsule, an hour before bed.
Liquid Serotonin - 5 to 10 drops, as needed, for agitation or insomnia.
DHEA - 1 (50 mg) capsule upon arising in the morning.
Pregnenolone - 1 (50 mg) capsule, upon arising in the morning.
Fortified Flax - For constipation, start with 1 teaspoon twice per day, dissolved in fruit juice. Increase up to 1 tablespoon, if needed; **OR Magnesium Chloride Liquid** - Use 1/2 to 1 teaspoon dissolved in fruit juice, once or twice per day, as needed for constipation.
Phosphatidylserine - 100 mg, three times per day, if cognitive functions are deteriorating.

Grief

The symptoms of grief are many, and the grieving process is a slow and painful process. The symptoms of grief behavior are extensive. These behaviors can be described under four general categories.

A) Feelings
B) Physical symptoms
C) Cognitive
D) Behaviors

A person in grief experiences a wide range of mental symptoms and feelings including depression, denial, anger, anxiety, fear, uncertainty, etc.

Physical Sensations Most Commonly Experienced

- Hollowness in the stomach
- Tightness in the chest
- Tightness in the throat
- Oversensitivity to noise, bright lights, and certain smells (such as hospitals).
- Unreality
- Breathlessness or shortness of breath
- Muscle weakness
- Fatigue
- Dry mouth
- Waking up several times nightly

Appetite disturbance presents a major problem and can cause long-term illness, if not addressed properly and corrected. In grief, your neurotransmitter level will be very low from the prolonged stress and anxiety. Grief can cause fuzzy or irrational thinking, as well as avoidance behaviors. Many people refuse to let go or "close the casket," and go on with their life. Staying in the past with a loved one can cause anxiety and phobias as well as physical illnesses.

Note: Find a therapist you can talk to and with whom you feel comfortable sharing your feelings. See him/her at least weekly through your acute stage, then on a monthly basis for the first year, and then as needed. Do not repress your feelings. Allow them to flow. If you took care of a loved one during a long illness, you can take on their symptoms which can include fear, depression, as well as their physical pain. Close the casket. Say good-bye, and let them go. Otherwise, your healing cannot begin.

Suggested Nutritional Support

Adults

Mood Sync - 1 or 2 capsules, twice to three times per day, in acute stage; and then 1 to 2, twice per day, as maintenance. Do not use if you are taking an SSRI antidepressant.

OR

Tyrosine* - 500 mg, 2 to 3 times daily, spread throughout the day.

B6 - 50 mg in the morning.

Brain Link Complex - 1 serving in the morning.

OR

TL-Vite - 1 in the morning with breakfast.

5-HTP - 1 (50 mg) capsule, 1 hour prior to bedtime. Do not use if you are taking an SSRI antidepressant.

Liquid Serotonin - Use 10 to 15 drops as needed throughout the day.

Mag Link - 2 tablets, twice daily.

Anxiety Control - For acute episodes of anxiety, 1 or 2 capsules twice to three times daily, divided, as needed.

Children

Brain Link Complex - use 1/2 scoop in juice, in the morning.

Tyrosine* - 200 to 500 mg depending on weight, daily. Under 50 pounds, use 200 mg.

Liquid Serotonin - Use 6 to 10 drops, as needed, throughout the day.

Anxiety Control 24 - 1 capsule, mid-morning, afternoon, and 30 minutes before bedtime.

Teenagers

Teen Link - Use 1 or 2 capsules, twice daily. Do not use if you are taking an SSRI antidepressant.

Brain Link Complex - 1 serving in the morning, and 1 in the evening.

**Do not use Tyrosine, Phenylalanine, or DLPA if you are pregnant or lactating, have PKU or if you use MAO inhibitors or tricyclic antidepressants, or if you have had a malignant melanoma.

- Avoid sugar.
- Avoid prolonged periods in dark rooms or rooms with drapes closed except when sleeping.
- Do not use alcohol for depression.

Grieving Process Those who cared for a sick relative are hit harder and longer by grief and depression. A study sponsored by the National Institute of Mental Health found 30% of caregivers suffer from clinical depression or anxiety while their loved one is alive. Four years later 25% suffered symptoms; only 10% of non-caregiving relatives were depressed for four years after death.

Grief from Pet Loss

Your pets became part of your family, and when its life comes to an end, you can experience deep feelings of grief and loss. There is no set time for your grief, and how long it will last. Don't be ashamed to express your feelings of loss. Your pet was intertwined with the daily rhythms of your life, especially when you first came home, at meals, and at bedtime. You experience the same deep, lost, and lonely feelings you do with a human loss. Don't suppress your feelings. Allow them to flow so you can put closure on the loss, and allow the healing process to begin. Use the nutritional support program outlined, as long as needed. For special help, contact the Delta Society at 1-800-869-6898; they are available 24 hours per day. They specialize in counseling those who have lost a pet and need assistance with their grief.

Headaches

Headache is the number-one complaint of the American public. Consumers spend approximately 6 billion dollars per year on pain medications and over-the-counter formulas for relief. It is estimated that yearly approximately 90% of men and 95% of women suffer from headaches painful enough to send them to doctors' offices. The source of the headaches is not located in the brain, itself, as there are no sensory nerves in the brain. Pain produced inside the skull is rare and usually due to tumors or other disorders; the pain is a secondary symptom. Most headache pain originates *outside* the skull in the nerves leading to the muscles and blood vessels around the face, scalp, and neck.

The most common types of headaches include migraine, tension or muscle contraction, and cluster. Often, a person can experience a combination of headaches. But headaches may also be caused by other underlying problems, such as sinus or allergy.

Food or stress can trigger migraines. Migraines are often called the avoidance headache. Migraine headaches produce a throbbing pain on one side of the head, but the pain can spread to the entire head. Often nausea and sometimes vomiting occur. Visual symptoms are common. Facial tingling or numbness may occur. Other symptoms include extreme sensitivity to noise and lights. Usually the attacks last from 4 to 72 hours without treatment, and commonly interfere with normal activity to some extent. Migraine sufferers may look pale and feel cold. Sometimes the victim gets a forewarning of an attack with malaise, fatigue, and mood changes. It is not uncommon for the sufferer to feel exhausted and mentally foggy for hours after an attack.

Headache triggers include:

- Stress, anxiety, anger, or depression.
- Menstruation, oral contraceptives, or hormone-replacement drugs.
- Foods such as dairy, MSG, eggs, anything pickled, alcohol (beer or red wine), coffee, teas, chocolate, wheat, cheese, or tomatoes are the most common. But any food can be a culprit. Explore food allergies as a contributing cause. There are lab tests for differentiating foods, or you can do a food-elimination diet.
- Environmental substances such as perfumes, paint, new carpet, glues, fumes, etc.
- Missed or delayed meals.

- Flickering fluorescent lights, sunlight.
- Time-zone changes.
- Holidays and travel.
- Strong smells.
- Loud noises.
- Bone structure misalignment and muscle spasms with trigger points.
- Alteration of sleep-wake cycle, such as sleep deprivation or excesses.
- Certain drugs.

Tension or muscle-contraction headaches are probably the most common form of headaches. Tension headaches account for 75% of all headaches, and are usually a response to stress, fatigue, or environmental factors; they can even start after a stressful event. The pain in the head results from muscle contractions of the head, neck, back, or facial muscles. Pain is often felt in the forehead, extending up from the base of the skull. The muscles of the upper back or neck contract, causing the pain to be *referred* to the head. This is frequently described as a tight feeling, like a band or vise around the head. The neck and shoulders often feel sore, and the person can develop trigger points in the muscles. Trigger points are sore, tender points that form scar tissue within the muscle. The trigger point causes the muscle to go into contraction from stress, anxiety, overuse, poor posture, or staying in same position for an extended period of time. The headaches can become chronic and occur daily. Headaches can last only a few minutes, but usually last several days. Nausea is uncommon with muscle-contraction headaches, and usually do not limit activities as do migraines. But muscle-contraction headaches can make you feel that you are going crazy from the pain. The chronic muscle-contraction headache can bring on depression, anxiety, and sleep problems. Massage with deep-tissue work sometimes benefits. But if the muscle does not let go, it may be necessary to have trigger point injections to break up a headache.

Cluster headaches cause very severe one-sided head pain. Usually the pain centers in the eye. Other symptoms include excessive tearing, drooping eyelid, stuffy or runny nose, all on the side of the pain. Restlessness is a common symptom. The pain can be so severe that the sufferer paces or bangs their head against the wall to contend with the headache. Cluster headache attacks usually last 30 to 90 minutes, but they can last hours. The sufferer generally has recurrent attacks over a month to three months, with the headaches occurring during an active phase once or twice daily, or every other day. Then the headaches do not recur for several months to years. There is no known cause for cluster headaches.

About 90% of headaches are due to the aforementioned. Other most common causes of headaches include the following.

- Sinus headache. Due to increased pressure in the sinus cavities; often a sinus infection. Sinus headaches last until the sinuses drain. A sinus infection should be treated with antibiotics.
- Temporomandibular Joint Dysfunction (TMJ). Usually occurs in the *temple,* ear, or cheek regions of the head. Caused by clenching of the jaws, or grinding of the teeth. (Usually at night or from abnormalities of the jaw joint itself. Stress intensifies TMJ pain).
- Glaucoma. Increased intraocular (eye) pressure. Acute glaucoma may cause a throbbing pain around or behind the eye, or in the forehead. Eye redness and vision of halos or rings around lights may occur with glaucoma.
- Hypertension. Increased blood pressure; can contribute to a headache.
- Strokes, aneurysms and brain hemorrhages. A severe headache of sudden onset associated with stupor, or other neurological symptoms, demands prompt medical evaluation.
- Head trauma can cause pain, and can reflect serious damage ranging from fractured skull to internal bleeding.
- Occipital neuralgia headache. Occurs mainly in senior citizens. Symptoms include jabbing pain in the back of the head and neck, tenderness in neck and shoulder area.
- Other miscellaneous causes of headache include eyestrain, allergies, dental problems, dehydration, systemic infections, caffeine withdrawal, meningitis, brain swelling, and intense physical exertion.

If you have persistent pain that does not respond to rest and treatment, consult a physician. *If you experience a headache accompanied by severe pain, drowsiness, confusion, mood swings, visual disturbances, weakness or paralysis, consult a physician immediately.*

Suggested Nutritional Support for Headaches

Powerelief* - 1 or 2 caps every 4 to 6 hours, as needed, for relief.

Powerelief combines DLPA, Boswella, GABA, Passion Flower, Magnesium, and B6.

DLPA 750* -1 or 2 capsules, twice per day.

Mobigesic - 1 tablet, three times per day.

Boswella (300 mg) - 1 or 2, twice per day. Boswella is a herb from India, used for inflammation or swelling.

Anxiety Control 24 - 2 capsules, twice to three times per day.

Mag Link - 4 to 6 tablets per day, divided. If loose stools occur, then decrease by 1.

Feverfew - 1 capsule, twice per day, for prevention of migraines or to extend the life of DLPA.

Alka-Selzer *Gold* - 2 tablets dissolved in a glass of water at onset of a migraine headache. Often, this brings immediate relief for migraine headaches triggered by allergic reactions to food or chemical substances. The Alka-Selzer *Gold* helps to neutralize the allergic mechanism, and prevents the migraine from becoming full blown.

Riboflavin - 400 mg, daily, for prevention or reduction of migraines.

Accuband Magnets - Use on trigger points on the neck and upper back to relieve pain. Accubands are tiny, powerful magnets about the size of the top of pencil lead. They are applied with a small adhesive patch. These tiny magnets begin to ease pain immediately.

B Complex Capsule - 1 capsule in the morning.

Mobisyl Creme - Apply to neck and back once or twice per day.

NAC - For sinus congestion/headache, 1 (600 mg) capsule, twice daily.

- Elevate head above waist.
- Apply icepack to base of neck and on upper shoulders for 20 minutes at a time, then off for 30 minutes. Repeat as needed. (Try the SofTouch icebag—comes in its own cover, ready to use, and reusable. SofTouch packs are both freezeable and microwavable.
- Avoid heating pad for muscle-contraction or migraine headaches.
- For sinus headache, apply hot, moist towels or apply specially designed moist heating pad to the sinuses to facilitate sinus drainage.
- Massage to neck and shoulders can help relieve headache.
- Consider medical evaluation and trigger point injections, if headache persists.

* Do not DLPA take if you are pregnant or lactating, have PKU, or if you use MAO inhibitors or tricyclic antidepressants, or if you have had a malignant melanoma.

Heart Disease

Heart disease or coronary heart disease (CHD) is the number one killer of Americans, and causes half the deaths in the U.S. today. An estimated 57 million Americans are afflicted with some type of heart disease. Heart disease claims over 720,000 people each year.
Risk factors include the following.

- Male gender, or post-menopausal women
- Genetics, history of heart disease in family
- Age
- High blood cholesterol
- Hypertension or high blood pressure
- Smoking
- Excess weight
- Lack of exercise
- Stress

Your sex, family history, and age pose risk factors beyond your control. But the other risk factors you *can* control. The most dangerous are high cholesterol, hypertension, and smoking. Each risk factor increases the risk of having a heart attack 2 to 3 fold, and all the risk factors compound each other dramatically.

Studies have shown that high cholesterol levels are directly proportional to your risk of heart attack. Diet is the next highest component that determines your cholesterol level. Dean Ornish, M.D., has proven cholesterol and your risk of heart disease can be dramatically changed with dietary and lifestyle changes, and love. Be aware, if you do not eat enough fat or cholesterol, the body increases the production of cholesterol in the liver. The body interpretes the reduction of fats or cholesterol in the diet to a time of famine or starvation. This causes insulin to activate HMG Co-A reductase, an enzyme in the liver, to manufacture more cholesterol than is required by the body from sugars and carbohydrates.

Cholesterol lowering drugs work by inhibiting the HMG Co-A. Many cholesterol-lowering medications also lower the CoEnzyme Q10 level and can cause liver damage. This can put you at higher risk for having a heart attack. CoQ10 is an enzyme necessary for cell oxygenation, and is essential for health of all tissues and organs in the body. As we age, the CoQ10 level drops dramatically. CoQ10 helps protect against heart attacks, relieves angina, boosts the immune system, lowers blood pressure, is an antioxidant, and helps periodontal disease.

Hypertension or high blood pressure afflicts 35 million, or 1 in 6 Americans. If hypertension is present, the likelihood of CAD is 3 to 5 times higher than if the person has normal blood pressure. It is important that you have your blood pressure checked by a healthcare professional, at least twice yearly.

Smoking is the third controllable risk factor. Quit! A complete nutritional program is outlined in our *Breaking Your Prescribed Addiction* book. Your risk of a heart attack one year after you quit smoking is just 10% greater than the non-smoker; after 5 years of abstinence, your risk factor becomes the same.

Suggested Nutritional Support

Carnitine - 1,000 to 3,000 mg per day, divided, for high cholesterol and triglycerides. Carnitine also increases the HDL (good) cholesterol. Immediately after a heart attack, supplement with 2 grams of carnitine to facilitate expansion of the heart muscle again. Continue carnitine at the 2 gram level. This helps reduce heart muscle damage while reducing angina and arrhythmias (abnormal heart rhythms).

Taurine - 1,000 mg twice, daily. Taurine is the most abundant amino acid in heart tissues. Taurine increases left ventricle function without changing the blood pressure, and helps balance the calcium and potassium in the heart.

Mag Link - 4 to 6 tablets per day, divided. If loose stools occur, decrease the dose by 1, or try spreading out the interval between doses. If you have CAD or hypertension, you are deficient in magnesium. Magnesium is nature's muscle relaxant, and is vital to a healthy heart. Mag Link is a magnesium chloride, the same form of magnesium found in the body; absorption and tolerance are best with this form of magnesium.

Chromium Picolinate - 200 mcg, per day. Chromium helps lower total cholesterol and triglycerides, while raising HDL cholesterol.

Vitamin E - 400 to 800 I.U., per day. Vitamin E is an antioxidant. Recent studies demonstrated a lower risk of fatal heart attack, if Vitamin E is taken daily.

DHEA - 25 to 50 mg per day, upon arising in the morning, on an empty stomach. (Have your physician check your DHEA sulfate level to obtain a starting point, and then have it checked 2 to 3 months after starting DHEA). In patients with CAD, the blood levels of DHEA have been very low; i.e., DHEA sulfate level of 4 (normal is 250—900).

CoEnzyme Q10 - 100 mg, per day, for CHD or hypertension. For congestive heart disease, increase CoEnzyme Q10 to 300 to 500 mg, per day, plus magnesium and B6.

Deluxe Scavenger - 3 per day divided. Deluxe Scavengers are a combination formula comprising CoEnzyme Q10, beta-carotene, Vitamin C, Lemon bioflavonoids, rutin, Vitamin E, Selenium, Glutathione, NAC, riboflavin, P5'P, and Vitamin B6.

Ester C - 2,000 to 5,000 mg per day, divided.

Garlic - 1,500 to 3,000 mg, divided. Garlic is an alternative to aspirin therapy. Ajoene, a component of garlic, is at least as potent as aspirin.

Folic acid - 400 to 800 I.U., daily. Folic acid is necessary for the proper metabolism of homocysteine. Excess homocysteine causes arterial plaque buildup.

Anxiety Control 24 - 1 or 2 capsules, twice or three times daily, as needed for anxiety or stress.

Rodex B6 - 150 mg capsule in the morning, for hypertension and congestive heart failure.

Hawthorn - 1 to 1.5 grams freeze-dried berries, three times daily. Hawthorn improves the circulation of blood to the heart by dilating blood vessels and relieving arterial spasms.

Herpes

Herpes has become a major social disease. Herpes attacks are characterized by clusters of clear, fluid-filled vesicles on the genitalia or face, accompanied by severe pain and itching. Once a person becomes infected with the herpes virus, rarely will the virus become extinct. The virus rests dormant in the body after the initial infection. When it reactivates as a result of stress, such as emotional upset, sunburn, etc., the virus produces an outbreak. Stress causes reemergence of the virus, and changes the balance of the amino acids arginine and lysine.

Keeping the balance of lysine to arginine at the right levels prevents replication of the herpes virus, and keeps it in check. Lysine is relatively easy to get in the diet, and most people consume ten times the minimum. However, vegetarians tend to have low lysine levels. A key to keeping herpes under control is to watch the ratio lysine-to-arginine foods. You must find the proper balance for you by trial and error. Avoid arginine-rich foods: chocolate, carob, coconut, oats, peanuts, soybeans, wheat germ, gelatin. Increase lysine foods: beef, chicken, lamb, milk, cheese, beans and brewer's yeast. In addition, take lysine supplementation in amounts of 500 to 1,500 mg per day.

Suggested Nutritional Support

Lysine - 500 to 1,500 mg per day as maintenance. During acute outbreaks, increase the lysine up to 2,000 mg, per day, and add another 1,000 mg of Vitamin C (Ester C).

Ester C with Bioflavonoids - 500 mg, three times per day.

Anxiety Control 24 - 1 or 2 capsules, twice per day. During an acute outbreak, increase to three times per day.

B Complex - 1 capsule, daily.

Lysine Cream - Apply to sores at onset, and repeat three to four times per day.

Powerelief*- 1 or 2 capsules, twice per day, as needed for pain. Powerelief contains DLPA, boswella, GABA, magnesium, and B6.

Vitamin E - 400 I.U. capsule, daily.

Zinc - 30 mg per day for acute outbreak until healed; then 15 mg per day for maintenance. Zinc is important to skin.

If you feel tingling on an area around your mouth, apply an ice cube to the area for 5 minutes, then off for 5 to 10 minutes; repeat several times daily. This interrupts the viruses replication cycle, and may prevent a blister formation.

*Do not use if you are pregnant or lactating, have PKU or if you use MAO inhibitors or tricyclic antidepressants, or if you have had a malignant melanoma.

Insomnia

NIH reports 18 to 20% of the American public suffers from insomnia. There are several forms of insomnia—the ability to fall asleep when you first go to bed; or constantly waking up during the night and being unable to go back to sleep. Waking up the early morning, unable to return to sleep, in most cases is due to anxiety. Insomnia can be the result of varied causes, including pain, anxiety, depression, grief, stress, fear, caffeine consumption, stimulant drug use, and certain psychoactive drugs.

Many women with PMS, or those approaching menopause, can experience interrupted sleep patterns due to hormonal imbalances. Millions of people suffer from a condition known as restless leg syndrome, or leg cramps. This condition caused by a deficiency of magnesium, does not allow the muscles to relax, so they twitch and jerk all night. Magnesium deficiency can cause insomnia. Sleep apnea is a major problem and should be addressed by a physician who specializes in sleep disorders. If you have a problem with insomnia, do not consume any caffeinated coffee or beverage after 2 P.M.

Suggested Nutritional Support

5-HTP - 1 (50 mg) capsule,* 30 minutes prior to bed to elevate your serotonin level. Increase to 2, if needed. Do not use if you are taking an SSRI antidepressant.

Mag Link - 2 tablets, twice to three times daily. Take to bowel tolerance, then decrease by one.

Melatonin - Start with a 2 mg capsule, 30 minutes prior to bedtime. Use melatonin *only* if you are over 30. If you experience weird dreams or nightmares, you are taking too large a dose.

Anxiety Control - 2 capsules, approximately 30 minutes prior to bedtime. If you wake up after a few hours, take an additional AC.

Sedaplus - 1 or 2 capsules, approximately 30 minutes prior to bedtime. Sedaplus is a herbal compound effective for sleep.

Powerelief* - 2 capsules, 30 minutes prior to bed if you suffer from pain.

Mobigesic - 1, as needed for pain, during the night.

Mobisyl - Rub on painful areas to reduce pain.

* Do not use Tyrosine, Phenylalanine, or DLPA if you are pregnant or lactating, have PKU or if you use MAO inhibitors or tricyclic antidepressants, or if you have had a malignant melanoma.

- Play a relaxation tape to help you relax and to promote sleep.
- Use deep breathing excercises for 10 to 15 minutes to relax the body and mind.

*Always use capsules. They breakdown more rapidly in the body, so you get to sleep more quickly. Capsules are generally purer, as they contain no binders.

Menopausal Stress/Anxiety

Many women experience anxiety just before menopause. Your body is undergoing a major transition; you must be patient and allow your brain and body to adjust to all the changes. Using amino acids for stress and anxiety will lessen the fear and uncertainty.

Suggested Nutritional Support

Gynovite - 2 tablets, three times daily. Gynovite is a multi-vitamin/mineral designed for postmenopausal women.

Mood Sync - 1 or 2 capsules, twice to three times daily.

B Complex 100 - 1 capsule, daily, in the morning with breakfast.

B.N.C. Plus GABA - 1 teaspoon in fruit juice in morning and the evening.

GABA 750 - 1/2 capsule in herbal tea or cup of water in the morning and mid afternoon.

5-HTP - 1 or 2 capsules, 30 minutes prior to bedtime.

Liquid Serotonin - Use 10 to 15 drops, under the tongue, twice daily, and at bedtime.

Carnitine - 500 mg twice daily, morning and afternoon.

Mag Link - 2 in the morning and 2 in the afternoon. If loose stools occur, decrease the dosage by one.

DHEA - 1 capsule daily, (25 mg or 50 mg, depending on age) upon arising in morning.

Pregnenolone - 1 (10 mg) capsule, daily.

If you are stressed out or anxious prior to meals, use liquid serotonin and 1/4 capsule of GABA in 8 ounces of warm water. Stress causes a release of insulin and adrenaline that tightens up the digestive muscles, causing bloating. If this is a chronic condition, consider digestive enzymes.

NOTE: For additional information on the dosage for DHEA, read *Health Educator Report* #51, and for Pregnenolone, read *Health Educator Report* #20.

*Do not use if you are taking an SSRI.

Obsessive-Compulsive Disorder

Obsessive-compulsive disorder (OCD) is characterized by anxious thoughts or rituals you feel you cannot control. If you have OCD, you can be plagued by persistent, unwelcome thoughts or images, and the urgent need to engage in certain rituals.

You may be obsessed with germs, or dirt, so you wash your hands over and over. You can be filled with doubt, and feel the need to check things again and again. You might be preoccupied by thoughts of violence, and fear that people close to you will be harmed. You may spend long periods of time touching things or counting. Or you may be preoccupied by order or symmetry. You may be worried by thoughts that are against your religious beliefs.

The disturbing thoughts or images are called obsessions. The rituals performed to prevent or dispel them are called compulsions. There is no pleasure in carrying out the rituals you are drawn to, only temporary relief from the discomfort caused by the obsession.

Many healthy people can identify with having some of the symptoms of OCD, such as checking the stove several times before leaving the house. But the disorder is usually diagnosed only when such activities consume at least an hour a day, prove very distressing, and interfere with daily life.

Most adults with this condition recognize the senselessness of their condition, but they cannot stop. Although some people, especially children with OCD, may not realize their behavior is unusual.

OCD strikes men and women in approximately equal numbers and afflicts roughly 1 in 50 people. It can appear in childhood, adolescence, or adulthood, but usually starts in the teens or early adulthood. A third of adults with OCD experience their first symptoms as children. The course of the disease varies. Symptoms can come and go. They may ease over time, or they can grow progressively worse. Evidence suggests that OCD might run in families.

Depression or other anxiety disorders can accompany OCD. Some people with OCD have eating disorders. Additionally, they avoid situations in which they might have to confront their obsessions. They may try unsuccessfully to use alcohol or drugs to calm themselves. If the OCD grows severe enough, it can keep someone from holding down a job, or from carrying out normal responsibilities at home. Frequently, however, the disorder does not develop to those extremes.

Research by National Institute of Mental Health scientists and other investigators led to the development of amino acids and behavioral treatments benefiting people with OCD. A combination of the treatments often benefits most patients. Some individuals respond best to one therapy, some to another. Behavioral therapy, specifically a type called *exposure and response prevention,* has also proven useful for treating OCD. It involves exposing the person to whatever triggers the problem, and then helping him or her forego the usual ritual. For instance, having the patient touch something dirty and then not washing his hands. This therapy succeeds in patients who complete a behavioral therapy program, although results have been less favorable in some people who have both OCD and depression. Address OCD and depression separately.

Suggested Nutritional Support

Mood Sync - 1 or 2 capsules, twice to three times daily. Mood Sync combines 5-HTP, St. John's Wort, GABA, glutamine, taurine, and B6. Do not take if you are taking an SSRI antidepressant.

OR

Tyrosine 850 - 1 in the morning, and 1 in the evening. If under 100 pounds, use tyrosine 500 mg, 1 in the morning and 1 in the evening. Do not take if you are taking a MAO inhibitor or tricyclic antidepressant.

TL-Vite - 1 capsule, in the morning.

Glutamine - 1,000 mg, three times per day.

GABA 750 - 1/2 capsule dissolved in water, mid morning and mid afternoon.

Liquid Serotonin - 10 to 15 drops, four times per day, or as needed.

5-HTP - 2 capsules, 1 hour before bedtime. Do not take if you are taking an SSRI antidepressant.

B.N.C. Plus - 1/2 teaspoon in fruit juice, mid morning and mid afternoon.

Mag Link - 1 tablet, three times per day.

B Complex Capsule - 1 capsule in the morning.

Methionine - 500 mg, morning and evening.

Panic Disorder

Panic disorder strikes at least 1.6% of the population, and is twice as common in women than in men. It can appear at any age, but most often it begins in young adults. Not everyone who experiences panic attacks will develop panic disorder. Many people have one attack, but never have another. If you do have panic disorder, it is important to seek treatment. If untreated, the disorder can become very disabling.

Panic Attack Symptoms:

Pounding heart
Chest pains
Lightheadedness or dizziness
Nausea or stomach problems
Hot flushes or chills
Tingling or numbness
Shaking or trembling
Feelings of unreality
Shortness of breath
Sensation of choking or smothering
Tremors
A feeling of being out of control or going crazy
Fear of dying
Sweating

People with panic disorder experience sudden feelings of terror, repeatedly with no warning. While impossible to predict when an attack will occur, many develop intense anxiety between episodes, worrying when and where the next one will strike. In between times there is a persistent, lingering worry that another attack could come any minute.

When a panic attack strikes, your heart pounds and you feel sweaty, weak, faint, or dizzy. Your hands tingle or feel numb, and you feel flushed, or chilled. You can have chest pain or smothering sensations, a sense of unreality, fear of impending doom, or loss of control. You genuinely believe you are having a heart attack or stroke, losing your mind, or on the verge of death. Attacks can occur any time, even during non-dream sleep. While most attacks average a couple of minutes, occasionally they can go on for up to 20 minutes.

Often accompanied by other conditions such as depression or alcoholism, panic disorder may spawn phobias which can develop in places or situations where panic attacks have occurred. If a panic attack strikes while you are riding an elevator you can develop a fear of elevators, and start avoiding them.

Many people's lives become greatly restricted. You avoid normal, everyday activities such as grocery shopping, driving, or even leaving your home. You may be able to confront a feared situation only if accompanied by a spouse or other trusted person. Basically, you avoid any situation you fear might make feel you helpless if a panic attack occurs. When a person's life becomes so restricted by the disorder, as happens in approximately one third of all people with panic disorder, the condition is called agoraphobia. A tendency toward panic disorder and agoraphobia runs in families. Often, early treatment of panic disorder can stop the progression to agoraphobia.

Studies have shown proper treatment called cognitive behavioral therapy, orthomolecular therapy, or a combination of the two, helps 70 to 90% of people with panic disorder. Significant improvement occurs within 8 to 10 days.

The cognitive behavioral approach teaches you how to view the panic situations differently, and demonstrates ways to reduce anxiety by using breathing exercises or techniques to refocus your attention. Another technique used in cognitive behavioral therapy is called exposure therapy. This frequently helps alleviate the phobias resulting from panic disorder. In exposure therapy, you are very slowly exposed to the fearful situation a number of times until you become desensitized.

Some people find the greatest relief from panic disorder symptoms when they use orthomolecular therapy. Orthomolecular and cognitive behavioral therapies, can help relieve panic attacks and reduce their frequency.

Suggested Nutritional Support

Panic and anxiety can cause feelings of fatigue and loss of appetite. Nutrients taken daily in the morning are readily absorbed and give you a needed lift.

Brain Link Complex - Take with 8 ounces of fruit juice, and follow directions on can (dosage depends on weight).

Anxiety Control 24 - 2 capsules, early morning, noon, and evening. Add 2 additional, if needed for panic.

OR

GABA 750 - 1/2 capsule dissolved in water, mid morning. Repeat mid afternoon and evening.

Liquid Serotonin - 10 to 15 drops, three times per day.

5-HTP - 1 (50 mg), an hour before bedtime. Do not take if you are taking an SSRI antidepressant.

Powerelief - 2 capsules, as needed for pain. (Do not use if you are taking MAO inhibitor or tricyclic antidepressant medications.)
Mag Link - 2 tabs, twice to three times daily. Titrate to bowel tolerance. If loose stool/diarrhea occurs, try spreading out further or decrease by 1.

Premenstrual Syndrome (PMS)

Premenstrual Syndrome, or PMS, is a very distressing syndrome for many women. It commonly occurs in women of child-bearing age, 7 to 10 days prior to onset of her period. The degree varies, but at least 50% of all women experience at least some symptoms of PMS. In about half of these, the symptoms manifests so severely they need treatment or medical attention.

There are various risk factors for PMS. Advancing age and the number of children increase the development of PMS. As a woman ages, the more painful periods experienced by women in their teens and twenties are often replaced by the appearance of PMS symptoms. The most severe and difficult cases of PMS are usually found in women in their forties. PMS symptoms usually start to overlap menopausal symptoms.

Categories of PMS Symptoms

PMS A	anxiety, nervous tension, mood swings, irritability
PMS B	weight gain, swelling of breasts and extremities, breast tenderness, abdominal bloating
PMS C	headache, increased appetite, cravings for sweets, fatigue, dizziness, fainting, pounding heart
PMS D	depression, forgetfulness, crying, confusion, and insomnia
Other	oily skin, acne, clumsiness, feelings of violence, or even suicide in severe cases

Suggested Nutritional Support

Rodex B6 (timed-release) - 1 capsule (150 mg) in the morning.

Anxiety Control 24 - For irritability and anxiety, use 2, three times per day.

Mood Sync - 1 or 2 capsules, twice to three times daily, as needed for mood swings and depression. Do not use if you are taking an SSRI.

Tyrosine 850** - For depression, 1 in the morning and 1 mid afternoon.

Glycine - 1 or 2 capsules dissolved under tongue to cut sugar cravings.

Mag Link - 4 to 6 tablets per day, divided. Take to bowel tolerance, then decrease by one. Magnesium is extremely important for many symptoms of PMS, including anxiety, painful menstruation, and headache prior to period. Be patient with painful menstruation as it may take several months to increase the magnesium level.

Powerelief** - For pain relief, use 1 or 2 capsules, as needed.

Optivite - 2 tablets, three times per day, with meals, is a combination multi-vitamin/mineral designed for PMS.

Glutamine - For memory and concentration, use 1 (500 mg) cap, or ½ scoop glutamine powder, three times per day.

Chromium Picolinate - For sugar cravings and to help stabilize blood sugar, use 1(200 mcg), twice per day.

5-HTP - For sleep and to balance the serotonin level, take 1 (50 mg) capsule 30 minutes before bedtime. Do not use if you are taking an SSRI.

Vitamin E - For breast tenderness, increase up to 2,000 I.U. a week before your period, then decrease and maintain with no more than 800 I.U. of Vitamin E per day.

Femgest Cream - 1/4 teaspoon applied to back of hand, once or twice per day, as needed.

- Avoid caffeine, colas, and sugar to decrease anxiety, insomnia, nervousness. Colas and caffeine cause a loss of magnesium which can amplify anxiety.
- Avoid coffee, chocolate, cola based sodas, and high intake of salt based foods, to decrease fluid retention and breast tenderness.
- Reduce tobacco intake.
- Reduce the intake of fats, avoid fried foods and animal fats.
- Increase intake of fiber with complex carbohydrates, leafy vegetables, legumes, and fruits.
- Limit intake of alcohol, refined sugar, and red meat.
- Reduce stress through regular exercise. Use other stress-reduction techniques such as massage, meditation, and relaxation tapes.

**Do not use if you are pregnant, lactating, taking an MAO inhibitor or tricyclic antidepressant, or have had a melanoma.

Post Traumatic Stress Disorder (PTSD)

Patients demonstrate symptoms of generalized anxiety and reoccurring flashbacks of traumatic episodes. PTSD usually occurs after a sever major stressor: fire, tornado, major auto accident, unexpected death, or military experience. Weeks or months after the traumatic episode occurs, flashbacks can begin and cause numerous physical symptoms such as sweating, rapid pulse, increased anxiety, panic, insomnia, and avoidance behaviors. Flashbacks can also cause a release of adrenaline. This sets stress reaction into high gear. Stress burns amino acids, and unless the brain is supplied with nourishment, it will continue to send anxiety-related messages that cause physical symptoms. A therapist experienced in PTSD and a good orthomolecular program will be helpful and aid the healing process.

PTSD is a debilitating condition that affects thousands of people. Often, people with PTSD have persistent frightening thoughts and memories of their ordeal, and feel emotionally numb. PTSD was once referred to as shell shock, or battle fatigue. It was first brought to public attention by war veterans. But, it can result from any number of traumatic incidents, serious accidents, natural disasters such as floods or earthquakes, or violent attacks such as mugging, rape, kidnapping, or torture. The event triggering the post trauma may be something that threatened the person's life, or the life of someone close to them, or it can be something they witnessed, such as the Oklahoma City bombing.

People with PTSD can relive the trauma in the form of nightmares, or disturbing recollections in which they experience the event over and over in their minds. They may also experience sleep problems, depression, feelings of detachment or numbness, or be easily startled. They may lose interest in things they once enjoyed, and have trouble showing affection. Sometimes, they feel more irritable or more aggressive than before the incident, and can even become violent. Memories that remind them of the incident can be very distressing, causing them to avoid certain places or situations that bring back to the event. Even anniversaries of the event are often very difficult.

PTSD can happen at any age. The disorder can be accompanied by anxiety, depression, or substance abuse. Symptoms may be mild or severe. People may become easily irritated or have violent outbursts. In severe cases they may have trouble working or socializing. In general, the symptoms seem to be worse if the event that triggered them was initiated by a person, such as a mugging or rape, as opposed to something like a flood.

Ordinary events can serve as reminders of the trauma and trigger flashbacks or intrusive memories. A flashback may make the person lose touch with reality and reenact the event for a period of seconds or hours, rarely days. A person having a flashback, which can come in the form of images, sounds, smells, or feelings, believes the traumatic incident is happening all over again.

Not every traumatized person gets full-blown PTSD, some don't experience PTSD at all. If the symptoms last more than a month, PTSD is the diagnosis. Symptoms usually occur within 3 months of the trauma, but the course of the illness varies. Some people recover within 6 months; others have symptoms that last much longer. The condition can become chronic. In some cases, PTSD may not arise until years after the traumatic event.

Amino acids can ease the symptoms of depression and sleep problems, but psychotherapy, especially cognitive behavioral therapy, is an integral part of treatment. Being exposed to a reminder of the trauma as part of the therapy, such as returning to the scene, sometimes helps. Support from family and friends can help speed recovery.

Antidepressants and tranquilizers only postpone healing and suppress symptoms. They do not restore the brain chemistry with needed nutrients.

Suggested Nutritional Support

B.N.C. Plus GABA - 1 teaspoon in fruit juice, morning and evening.

Liquid Serotonin - 10 to 12 drops, 2 to 4 times daily, and if you awaken during the night.

Rodex B6 - 1 (150 mg timed release) capsule in the morning.

TL Vite (multi-vitamin) - 1 capsule, in the morning with breakfast.

Mag Link - 4 to 6 tablets per day, divided, throughout the day, up to bowel tolerance, then back decrease by one.

Anxiety Control24 - 2 capsules, three times daily, for the first month after a traumatic episode, then 2 in the morning and 2 in the afternoon. Increase to 6 per day, if needed.

Ester C - 1,000 mg, morning and evening.

Powerelief* - 2 capsules (300 mg), twice daily, as needed for pain. As an alternative, use Mobigesic, boswella, or DLPA.

Mood Sync** - If depression occurs, take 2 capsules, twice daily.

5-HTP** - 1 (50 mg) at bedtime. Take 2, if your sleep pattern is interrupted.

OR

Use **Melatonin** - For sleep use 1 (3 mg), an hour before bedtime.

- Do not use alcohol in the acute stage of trauma.
- Decrease caffeine intake, and have none after noon.
- Limit sugar and soda intake.
- During the first month, avoid loud noises and bright lights.
- Use relaxation tapes, two to three times per week.

Post-Trauma Case History

Kelly is a 36-year old college professor who never had any major problems with anxiety or chronic pain until May 15, 1994. That day, while driving to the University, Kelly's life took a turn for the worse. She was hit broadside by a teenager. Her car was totaled, and so was her life.

Kelly spent a week in the hospital for painful neck and back injuries. Her healing was slow, and she had to learn to live with chronic pain and certain restrictions. But each day, Kelly improved a little more and tried to put her life back in order.

A few months passed without any particular problems. Then one day while driving to her classes, she heard a sound that she had hoped she would never hear again ... A car, unable to stop, slammed into the car in front of her. Kelly was not hurt, but she was so full of fear she had a full-blown panic attack. This was the beginning of her post-traumatic stress disorder, and all the symptoms that go with it. As time went on, she began having flashbacks of her accident, and with the flashbacks came the physical symptoms. Kelly developed a dread of driving because she never knew when the flashbacks would occur. Her physical symptoms included palpitations, weakness, trembling, apprehensiveness, sweating, headaches, muscle spasms, and fear of dying. Kelly came in to see me. I explained post traumatic stress disorder and

* Do not use if you have PKU, have had a melanoma, are pregnant or lactating, or if you use MAO inhibitors or tricyclic antidepressants.

** Do not use if you are taking an S.S.R.I.

the symptoms. I started her on a relaxation program and nutritional supplements. She was totally depleted of neurotransmitters because of the chronic stress. Kelly's nutritional support program included Anxiety Control, Mag Link, Powerelief, and B-complex. Within a few weeks she showed a marked improvement. Her headaches stopped, and her driving anxiety diminished. Post-traumatic stress disorder can cause numerous psychological and physical problems. But with proper therapy healing will occur.

Deanol and Brain Function

Deanol is a natural precursor to acetylcholine. Adequate acetylcholine is necessary for proper brain, memory, and nervous system functioning since acetylcholine is the key neurotransmitter connection in the brain and skeletal muscle. However, age, medications, and illness all reduce the body's natural production of this vital substance.

According to the late Dr. Carl Pfeiffer, deanol is a provitamin, the body's building block for acetylcholine. Most supplements, such as choline and lecithin, taken to stimulate acetylcholine proves inefficient since they do not readily cross the blood-brain barrier; but deanol, taken sublingually, readily does.

Research studies indicate increased acetylcholine increases concentration, helps memory retention, calms the nerves, elevates the mood, increases muscle tone, and causes changes in sleeping habits (sounder sleep and clearer mind upon awakening). Since the mid 1970s, researchers have known Alzheimer's disease causes a deficiency of acetylcholine.

Pure deanol is the only natural biological stimulant that does *not* adversely affect blood pressure or sleep. It is the ideal supplement for people who have to maintain an active, attentive concentration for prolonged periods or to offset drowsiness caused by food, drink or medication.

Deanol is available in liquid form, and can be used by children and adults.

Amino Acid Testing

The body constantly conducts many complicated series of chemical reactions in precisely controlled ways to keep us healthy. Over 5,000 reactions occur every second in a cell. By utilizing the natural substances in optimal quantities to reestablish a normal balance, you can help correct the cause of some disease processes in a nontoxic way.

Amino acid metabolism disorders are becoming recognized as a major factor in many disease processes. Amino acid analysis is an analytical technique on the leading edge of nutritional biochemical medicine. It gives a new approach toward illness, and can assist patients who have not responded to treatment as expected, or who present complex cases with diverse symptoms.

Amino acid analysis of urine or plasma goes a long way toward assessing vitamin and mineral status. Amino acid analysis measures the levels of amino acids in the body that affect many important processes. Additionally, it provides insights into the patient's functional needs for a wide variety of vitamins and minerals. Many of the enzymes which catalyze the interconversion of amino acids require vitamin and mineral cofactors to function optimally. In many cases where incomplete conversion of one product to another is due to sluggish enzymes, this indicates a functional need for increasing levels of a cofactor.

Amino Acid Analysis has proven helpful in treating:

- Chronic Fatigue
- Candida Infections
- Food & Chemical Sensitivities
- Immune System Disorders
- Anxiety
- Depression
- Learning Disorders
- A.D.D./ Hyperactivity
- Behavioral Disorders
- Eating Disorders
- Cancer
- Hypoglycemia
- Diabetes
- Cardiovascular Disease
- Seizures
- Headaches
- Arthritis
- Chronic Pain

For detailed information, call the
Pain & Stress Center at (210) 614-7246.

General Summary of Amino Acids

In order to absorb and assimilate *any amino acid*, you must take B6 or P5'P per day. The other important cofactor is magnesium. Magnesium is a facilitator, and is involved in over 300 enzyme reactions in the body.

ALANINE

- May be used as a source for the production of glucose.

Therapeutic dosages: 200 to 600 mg per day.

ARGININE

- Helpful in treatment of burns, elevated ammonia levels, and cirrhosis of the liver by detoxifying ammonia.
- Stimulates immune response, by enhancing the production of T cells.
- Induces growth hormone release from the pituitary gland.
- Use with caution in schizophrenia.
- Keep intake low for herpes simplex and Epstein Barr virus interactions.
- Required for normal sperm count.
- Provides essential amino acid for children, not adults.
- Enhances fat metabolism.

Therapeutic dosages: Up to 3,000 mg per day.

ASPARTIC ACID

- Helps protect the liver.
- Promotes mineral uptake in intestinal tract.
- Helps in ammonia detoxification.
- Acts to transport magnesium and potassium to the cells.

CARNITINE

- Helps clear blood of triglycerides.
- Helps lower cholesterol, and increase HDL levels.
- Essential in transportation of long chain fatty acids into the cells where fats are converted into energy.
- In heart patients, increases exercise endurance.
- Increases muscle strength.
- In uremia or kidney disease, may reduce risk factor for atherosclerosis and coronary heart disease.
- Helps alleviate depression.

Therapeutic dosages: 1,000 to 3,000 mg per day.

CITRULLINE

- Produces the amino acids Arginine and Ornithine.
- Detoxifies ammonia or other "nitrogen" waste products.
- Stimulates growth hormone production.

CYSTEINE

- Derives from serine and methionine in the liver.
- Contains sulfur.
- Helps maintain skin flexibility and texture by slowing abnormal cross linking of collagen.
- Helps reduce free radical formation, acts as an antioxidant.
- Reduces iron deficiency, anemia, and helps promote of iron absorption.
- Promotes red and white blood cell reproduction and tissue restoration in lung diseases.
- Provides strength of the hair (over 10% of hair is cysteine).
- Converts to cystine, if Vitamin C is deficient.
- Chelates heavy metal in the body.
- Helps protect against effects of alcohol and pollution.
- Assists in prevention of cataracts.
- NAC is helpful with asthma, bronchitis and sinus drainage by liquefying and thinning mucus.
- NAC helps in preventing side effects from chemo and radiation therapies.

Note: Use with caution in diabetic, due to changes in insulin cycle.

Therapeutic dosages: 250 to 500 mg, twice per day, divided.

CYSTINE

- Helps reduce psoriasis and eczema.
- Enhances tissue recovery after surgery.
- Functions as part of insulin molecule.
- Helps keep hair healthy, high content found in hair.

Note: Use with caution in people predisposed to stone formation in the liver or kidneys.

GABA

- Reduces anxiety/stress by decreasing limbic firing.
- Helps decrease muscular tension due to stress.
- Assists with hyperactivity and A.D.D.
- Crosses blood-brain barrier.

Therapeutic dosages: 600 to 3,000 mg, per day, divided.

GLUTAMIC ACID

- Acts as precursor to GABA.
- Functions as component of glucose tolerance factor.
- Helps diminish muscular dystrophy symptoms.
- Helps maintain the body's nitrogen balance.

GLUTAMINE

- Helps heal ulcers and intestines.
- Helps protect body from effects of alcohol.
- Assists in treatment of alcoholism by decreasing the desire to drink.
- Crosses blood-brain barrier.
- Aids mental functions, such as memory and concentration.
- Helps reduce intestinal permeability with food allergies.
- Produces other nonessential amino acids.
- Enhances effectiveness of chemotherapy and radiation treatments for cancer, while reducing toxicity and damage to the body.

Therapeutic dosages: 500 to 4,000 mg per day, divided.

GLUTATHIONE

- Contains cysteine, glycine, glutamate.
- Helps reduce free-radical formation.
- Neutralizes atmospheric substances such as petrochemicals and chlorine.
- Protects against radiation therapy.
- Transports other amino acids into the cell.

Therapeutic dosages: 1,000 to 3,000 mg per day, divided.

GLYCINE

- Simplest amino acid.
- Has sweet taste and readily dissolves in liquids.
- Inhibitory neurotransmitter, (inhibits excitatory messages).
- Helps reduce anxiety and hyperactivity.
- Helps to remove lead from the body.
- Assists with epilepsy.
- Shows low levels in ALS.

Therapeutic dosages: 500 to 3,000 mg per day, divided.

HISTIDINE

- Provides essential amino acids for infants, not adults.
- Helps relieve pain due to rheumatoid arthritis.
- Always take with Vitamin C (Ester C).
- Use with caution in females prone to depression due to PMS.
- Provides mild anti-inflammatory effect.
- Metabolizes into the neurotransmitter histamine (involved in smooth-muscle function in blood vessels).
- Helps maintain the myelin sheath, or insulator, of certain nerves.
- Provides necessary nutrient for proper functioning of auditory nerves.
- Chelates toxic metals from the body.
- Helps alleviate nausea during pregnancy.

Therapeutic dosages: 1,000 to 6,000 mg per day.

ISOLEUCINE

- One of the Branch Chain Amino Acids. Take as a *group.*

Therapeutic dosages: 250 to 700 mg, per day. Best as BCAA group supplement.

LEUCINE

- Belongs to the Branch Chain Amino Acids. Take as a *group.*
- Produces energy under many kinds of stress—from trauma, to surgery, infection, muscle training and weight lifting.
- Substitutes for glucose while fasting (the only amino acid that can do this).
- Stimulates insulin release which stimulates protein synthesis and inhibits protein breakdown.
- Helps reduce symptoms of Parkinson's disease in large doses.
- Use with all stress.

Therapeutic dosages: BCAA as a group, use 175 to 1,200 mg per day.

LYSINE

- Found in muscle, connective tissue, and collagen.
- Inhibits growth and replication of herpes and Epstein Barr viruses.
- Promotes growth, especially bone growth in infants and children; both requires much larger amounts of lysine than adults.
- Is required for antibody formation.
- Tends to be low in vegetarians.

Therapeutic dosages: 1,000 to 3,000 mg per day (with Vitamin C), divided.

METHIONINE

- Belongs to one group of the sulfur amino acids.
- Acts as powerful antioxidant, and detoxifies the liver.
- Gives rise to taurine.
- Helps detoxify heavy metals from the body and remove excessive levels of histamine.
- Participates in synthesis of choline, adrenaline, lecithin, and Vitamin B12.
- Inhibits synthesis of homocysteine which promotes plaque deposits in arteries. (must be taken with B6).
- Provides necessary formation of selenium to the body.

Therapeutic dosages: 800 to 3,000 mg per day, divided.

ORNITHINE

- May reduce fat and increase muscle mass.
- Powerfully stimulates growth hormone production by pituitary gland.
- Assists detoxification of ammonia in urea cycle.
- Stimulates immune system and enhances wound healing.
- May be useful in autoimmune disease such as arthritis.
- Converts (in body) to arginine, glutamine or proline.
- Causes insomnia in some people at doses of 1000 mg or more.

Therapeutic dosages: 500 to 3,000 mg per day, divided.

PHENYLALANINE

- Precursor to catecholamines—epinephrine, norepinephrine, dopamine, and dopa.
- May help appetite control appetite by stimulating CCK (chaleocephtokinin) enzyme.
- Increases blood pressure in hypotension.

Therapeutic dosages: 500 to 2,000 mg per day, divided.

Note: Do not be take if pregnant, lactating, taking MAO inhibitors or tricyclic antidepressants, or if you have a history of melanoma.

PROLINE

- Helps in lower blood pressure.
- Helps repair muscle, tendon and collagen injuries.
- Assists in skin flexibility in relation to aging and sun exposure, and is essential for skin health.
- Exists as major amino acid in collagen, if Vitamin C is present.

Therapeutic dosages: 500 to 1,000 mg per day, with Vitamin C.

TAURINE

- Manufactured in the liver from methionine or cysteine, with B6. Zinc must be present for taurine to function properly.
- Found in foods of animal origin.
- Belongs to group of sulfur amino acids.
- Found in heart and assists in balancing calcium and potassium in the heart. Taurine is the predominant amino acid in the heart.

- Used more by the body when under stress.
- Eases physical signs and symptoms without side effects, for congestive heart failure patients.
- Increases left ventricular heart performance without changes in blood pressure.
- Associated with retinal (eye) degeneration.
- Helps reduce seizures and epilepsy.
- Detoxifies secondary bile acids and toxins.
- Protects cell membranes.
- Frequently considered a neuromodulator.
- Zinc enhances taurine's effect.

Therapeutic dosages: 500 to 3,000 mg per day, divided.

THREONINE

- Provides necessary nutrients for digestive and intestinal function.
- Fills requirement for an essential amino acid which rises three times normal during pregnancy.
- Prevents accumulation of fat in liver.
- Increases glycine levels in brain and notably reduces ALS symptoms.
- Provides nutrients essential for mental health.

Therapeutic dosages: 100 to 500 mg, per day.

TRYPTOPHAN (Use 5-HTP or 5-hydroxytryptophan)

- Functions as an essential amino acid and neurotransmitter, precursor to serotonin.
- Found in organs throughout the body (as neurohormone).
- Effective in treating pellagra, and is precursor to niacin.
- Appears to assist in blood clotting mechanisms.
- Helps alleviate insomnia, in doses of 50 to 100 mg.
- Helps to reduce depression, schizophrenia, and assists weight control.
- Helps with pain by elevating the pain threshold.
- Acts as mood stabilizer, calms aggression and obsessive behavior.
- Take with B6 with carbohydrates such as fruit juice for maximum uptake to brain.

- Has helped decrease tremors in Parkinson's patients.
- Currently unavailable in U.S. due to F.D.A regulations.

Note : Do not take with S.S.R.I.s drugs such as Prozac, Paxil, Luvox, or Effexor.

Therapeutic dosages: 5-HTP 50 to 100 mg day divided.

Tryptophan → 5-HTP → Serotonin

5-HTP is about 10 times stronger than tryptophan, and is only 1 step biochemically, away from serotonin.

Normal tryptophan dose is 500 to 4,000 mg per day, divided.

TYROSINE

- Helps alleviate symptoms of Parkinson's disease.
- Functions as precursor to catecholamines—epinephrine, norepinephrine, dopamine and dopa.
- Precursor of thyroid hormones.
- Assists in normal brain function and supplies neurotransmitters.
- Can be used in place of many antidepressants.
- Helps stabilize blood pressure.
- Involved in tissue pigmentation.
- Called the *Stress* amino acid.

Note: Should not be taken with MAO and tricyclic antidepressants, and when cancerous melanoma is present.

Therapeutic dosages: 500 to 2,000 mg per day, divided.

VALINE

- Third of the Branch Chain Amino Acids, and should be taken as a group; functions as an essential amino acid needed for the nitrogen balance in the body.
- Necessary for muscle coordination, and mental and neural function.
- Helps reduce inflammation.

Therapeutic Dosage: up to 1,000 mg daily. Best if taken as BCAA.

Magnesium/Amino Acid Connection

Symptoms of Magnesium *Deficiency*

Anxiety
Panic attacks
Mitral valve prolapse
Hypertension
Chronic pain
Back and neck pain
Muscle spasms
Migraines
Fibromyalgia
Spastic symptoms
Chronic bronchitis, emphysema
Vertigo (dizziness)
Confusion
Depression
Psychosis
Noise sensitivity
Ringing in the ears
Irritable-bowel syndrome
Cardiovascular disease
Cardiac arrhythmias
Atherosclerosis/Intermittent claudication
Raynaud's disease (cold hands and feet)
TIA's (Transient Ischemic Attacks-strokes)
Constipation
Fatigue
Diabetes
Hypoglycemia
Asthma
Seizures
Kidney stones
Premenstrual syndrome
Menstrual cramps
Osteoporosis

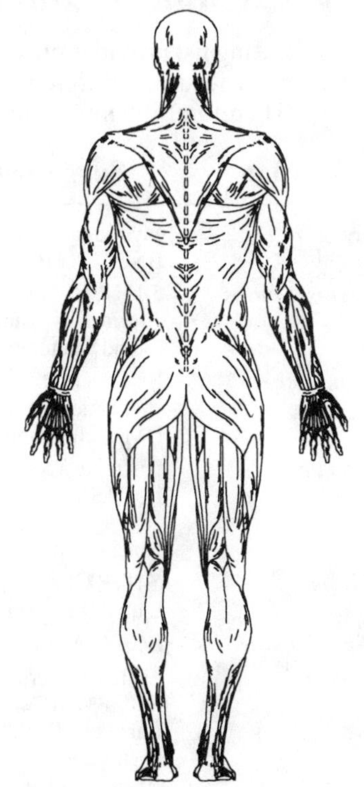

You have 657 muscles that need magnesium every second of every day. Magnesium is a cofactor for all amino acids. For maximum benefit, add some magnesium in the form of magnesium chloride such as Mag Link. If after adding GABA and other amino acids to your regimen, you still feel anxious, your problem may be magnesium deficiency. Add two Mag Link tablets morning and afternoon. You should begin to feel a decrease in symptoms.

Warning: People with renal or kidney failure should not take magnesium without medical supervision.

For detailed information on magnesium-deficiency symptoms, read *The Anxiety Epidemic.*

Amino Acids and Nutrients For Clinical Conditions and Diseases

The following list provides only a guideline for specific conditions. Each individual's needs are different. Some nutrients will help you more than others. Therefore, you must find the nutrients that best suit your situation.

Condition	Suggested Therapy	Avoid
Aging	Methionine, 5-HTP, Glutamine, Ester C, BNC + GABA, Melatonin, Pyncogenol	
Aggressiveness	5-HTP, GABA, Glycine, Glutamine, Taurine, Tyrosine, B6, BNC + GABA, Liquid Serotonin	Phenylalanine
Alzheimer's	BNC + GABA, Glutamine, Ginkgo, B6 (Timed-Release), Pycnogenol, DMG, Phosphatidylserine	
Arthritis	Histidine, Cysteine, BNC + GABA, Boswella, Glucosamine, Malic Acid +, DLPA, Ester C, Magnesium (Mag Link), Niacinamide, Shark Cartilage	
Autism	5-HTP, Glutamine, B6, Taurine, Liquid Serotonin, Magnesium	
Body Building	BCAA, Alanine, Carnitine	
Cancer	Cysteine, Taurine, Glutamine BNC+ GABA, Ester C, Shark Cartilage, BCAA, Melatonin, Pycnogenol	With Melanoma Phenylalanine, Tyrosine
Cholesterol	Carnitine, Methionine, Arginine, Glycine, Taurine, Fortified Flax, Chromium Picolinate	
Chronic Illness	BCAA, BNC + GABA, Cysteine, 5-HTP, Ester C, Pycnogenol	
Chronic Pain	5-HTP, DLPA, GABA, Glutamine, BNC + GABA, Boswella, Ester C, Mag Link (magnesium), Powerelief	
Cirrhosis / G.I. Healing	Glutamine, Carnitine, BCAA, CoEnzyme Q10	
Depression	5-HTP, Phenylalanine, Tyrosine, Methionine, GABA, Carnitine, Threonine, Taurine	Arginine
Diabetes	5-HTP, Carnitine, Taurine, Chromium Picolinate, Mag Link (magnesium), Pycnogenol, Ester C, Vitamin E	
Drug Addiction	GABA, Methionine, Tyrosine, Glutamine, DLPA, B Complex 5-HTP, B6 (Timed Release)	Alcohol
Energy	Carnitine, Tyrosine, CoEnzyme Q10	

Epilepsy	Glycine, Taurine, GABA, B6, Melatonin	
Gallbladder	Methionine, Taurine, Glycine, BCAA	
Gout	Glycine	
Hair Loss	Cysteine, Arginine	
Heart Failure	Taurine, Tyrosine, Carnitine, BCAA, CoEnzyme Q10, Mag Link (magnesium)	
Herpes	Lysine, Ester C, Zinc	
Hyperactivity	GABA, Glycine, Glutamine, Taurine, 5-HTP, Liquid Serotonin, BNC + GABA	
Hypertension	5-HTP, GABA, Taurine, Mag Link (magnesium), Calcium, CoEnzyme Q10	
Hypoglycemia	Alanine, GABA, Chromium Picolinate, Vanadium, Mag Link (magnesium)	
Insomnia	5-HTP, Melatonin, Liquid Serotonin, GABA, Mood Sync	Phenylalanine
Kidney Failure	BNC + GABA, Carnitine	
Leg Ulcers	Topical Cysteine, Glycine, Threonine, BCAA, Ester C, Zinc, Mag Link (magnesium)	
Liver Disease	BCAA, Carnitine, Glutamine, B6	
Manic	5-HTP, GABA, Glycine, BNC + GABA, Glutamine	Phenylalanine
Memory/ Concentration	Glutamine, GABA, Ginkgo, BNC + GABA, B6 (Timed Release)	
Mental Alertness	Tyrosine, Phenylalanine, Glutamine, Ginkgo	
Parkinson's	Phenylalanine, Tyrosine, 5-HTP, Methionine, Mag Link (magnesium), Ester C	
Radiation	Cysteine, Glutamine, Taurine, Ester C, Beta Carotene, Pycnogenol	
Schizophrenia	GABA, Isoleucine, 5-HTP, Methionine, B6, NonFlush Niacin	Serine, Leucine Aspargine
Seizures/Tics	Taurine, GABA, Mag Link (magnesium)	
Stress	Anxiety Control, Tyrosine, GABA, Histidine, BNC+, GABA,Glutamine, B6, Glycine, Ester C, Mag Link (magnesium)	
Surgery	BCAA, BNC + GABA, Glutamine, Ester C, Beta Carotene	
Tardive Dyskinesia	GABA, Taurine, BCAA, Glutamine Mag Link (magnesium)	
Tobacco Addiction	Tyrosine, GABA, 5-HTP, Glutamine Methionine, B Complex, Sulfonil	
Weight Control	5-HTP, Phenylalanine, GABA, Tyrosine, HCA	

Amino acids are involved in many metabolic pathways in the body. They are extremely important as detoxifying and immune-stimulating agents. Detoxifying amino acids include cysteine, glutamine, glycine, methionine, taurine, and tyrosine. Immuno-stimulating amino acids include: alanine, aspartic acid, cysteine, glycine, lysine, and threonine.

Drug-Nutrient Actions

Drug/ Condition	Parallel Nutrient	Opposite Nutrient
Anticonvulsants	Taurine, GABA, Glycine, 5-HTP, Magnesium	Aspartic Acid
Antidepressants	Phenylalanine, Tyrosine, Methionine,Taurine, St. John's Wort	Glycine Histidine
Heart Failure	Taurine, CoEnzyme Q10, Carnitine, Magnesium	Niacin, 5-HTP
High Cholesterol / Triglycerides	Carnitine, Magnesium, Chromium Picolinate	
Steroids (anabolic)	BCAA, Carnitine	Glutamic Acid
Viral antagonists	Lysine, Zinc	Arginine

Product Information

The *purity* of amino acids and nutritional supplements is very important to your success. Be selective. Your body responds to what you absorb. Absorption enhances when you *use only pharmaceutical-grade products.* There are four grades of supplemental amino acids and nutritional products. We list them in order of purity from least pure to purest:

1) Food lot
2) Cosmetic
3) Pharmaceutical grade
4) I.V. grade

Pharmaceutical grade generally guarantees purity of product. Capsules are generally cleaner and purer than tablets, and *more bioavailable.* Tablets require fillers and binders. When considering nutritional supplements, look for preservative-free and excipient-free products because they are better. Buy supplements that are free of preservatives, fillers, binders, or excipients of any kind. *Insist on pharmaceutical-grade products.* Your body and brain will know the difference.

For information on products and other books, call 1-800-669-2256.

Stress/Anxiety Reactions

Requiring More Amino Acids

Within 24 to 48 hous after a stress-anxiety-anger, or emotional-upset reaction, major physical symptoms *can and do* occur.

Headaches
Face / Body Pains
Neck / Back Pain
Trigger Points
Sleep Loss
Upset Stomach
Diarrhea
Constipation
Pounding Heart
Increased Sweating
Increased Anxiety
Jaw Clenching
Skin Eruptions
Elevated B.P.
Bladder Infections
Ulcers

The three categories of stress—emotional, chemical, and physical—lead to physiological changes in various organs. These changes cause a total modification and weakening of the immune system. If all three stress factors are combined and prolonged, a disease state may follow.

The importance of a balanced amino acid program proves vitally important. You must restore your neurotransmitters to relieve the emotional, chemical, and physical symptoms.

Bibliography

Aatron Medical Services. Inc. *Amino Acids Metabolism and Analysis.* 1989.

Amaducci L., et al. "Use of phosphatidylserine in Alzheimer's Disease." Annuals New York Academy of Science. Vol. 640, 1991, pp. 245–249.

Babal, Ken. "The Fall and Rise of Tryptophan," *Nutrition Science News,* February, 1998, Vol. 3, No. 2, pp. 60–64.

Balch, James F. and Phyllis A. Balch. *Prescription for Nutritional Healing.* Garden City Park, NY: Avery Publishing Group, 1997.

Bland, Jeffrey, Ed. *Medical Applications of Clinical Nutrition.* New Canaan, CT: Keats Publishing, 1983.

Bland, Jeffrey S. Psychoneuro-Nutritional Medicine: An Advancing Paradigm. *Alternative Therapies.* May, 1995, pp. 22–27.

Block, K.O. and In Friedman M. *Absorption and Utilization of Amino Acids.* Boca Raton: CRC Press, Volume 1, 1989.

Bliznakov, and Gerald L. Hunt. *The Miracle Nutrient: CoEnzyme Q10.* New York: Bantam Books, 1987, pp. 65–123.

Bowery, N.G., et al. (ed). *$GABA_B$ Receptors in Mammalian Function.* New York: John Wiley & Sons, 1990.

Breggin, Peter. *Talking Back to Ritalin.* Monroe, ME: Common Courage Press, 1998.

Breggin, Peter. *Toxic Psychiatry.* New York: St. Martin's Press, 1991.

Braverman, Eric and Carl C. Pfeiffer. *The Healing Nutrients Within.* New Canaan, CT: Keats Publishing, Inc., 1987.

Cenacchi T. et al. "Cognitive decline in the elderly: A double-bind, placebo-controlled muticenter study on efficacy of phosphatidylserine administration." *Aging Clinical Experimental Research.* Vol. 5. 1993, pp. 123–133.

Chaitow, Leon. *Thorsons Guide to Amino Acids.* London: Thorsons, 1991.

Chen, L.H. "Biomedical Influences on Nutrition of the Elderly." *Nutritional Aspects of Aging.* Boca Raton: CRC Press, 1986.

Cooper, Jack R., et al. *The Biochemical Basis of Neuropharmacology* New York: Oxford University Press, 1990.

Crook, T.H., et al. "Effects of Phosphatidylserine in Age-Associated Memory Impairment." *Neurology,* Vol. 41, 1991, pp. 644–649.

Crook, T, et al. "Effects of phosphatidylserine in Alzheimer's disease." *Psychopharmacolgy Bulletin.* Vol. 28, 1992, pp. 61–66.

Cross, L.D. "Carpal Tunnel Syndrome, Epidemic of the 90s." *Health Naturally.* April/May, 1998, pp. 34–37.

Davies, Stephen and Alan Stewart. *Nutritional Medicine (Vol 1).* New York: Avon Books, 1987.

Davis, Joel. *Endorphins, New Wave of Brain Chemistry.* New York, NY: The Dial Press, 1984.

Delong and C. Kagan. *Dematologia.* No. 156, pp. 257–267, 1978.

Dodson W., Sach D., Krauss S., et al. "Alterations of serum and urinary carnitine profiles in cancer patients. Hypothesis of possible significance. *Journal Ameican College of Nutrition.* Vol. 8, 1989, pp. 133–142.

"Four Prespectives on Carpal Tunnel Syndrome," *Your Health.* Asheville, NC: International Academy of Nutrition and Preventive Medicine. Vol. XIII No. III, 1992.

Fox, Arnold and Barry Fox. *DLPA To End Chronic Pain and Depression.* New York: Long Shadow Books, 1985.

Gitlin, Michael. *The Psychotherapist's Guide to Psychopharmacology.* New York: The Free Press, 1990.

Goldberg, Burton, et al. *Alternative Medicine Guide to Heart Disease.* Tiburon, CA: Future Medicine Publishing. 1998.

Granat, Q. and J. DiMichele. "Phosphatidylserine in Elderly Patients. An Open Trial." *Clinical Trials Journal.* Vol. 24, 1987, pp.99–103.

Griffith, R, D.Delong, and C. Kagan. Dermatologia. No. 156, pp. 257–267, 1978.

Harrison, Tinsley R., Thorn, George W. (ed), et al. *Harrison's Principles of Internal Medicine.* 8th Ed. New York: McGraw-Hill Book Co, 1977.

Hoffer, Abram and Morton Walker. *Orthomolecular Nutrition.* New Canaan, CT: Keats Publishing, 1978.

Hosking G.P., Cavanaugh N.P., Symth D. P., et al. "Oral treatment of Carnitine Mypopathy." *Lancet.* Vol. 16, 1977, p. 853.

Isaacs H., Heffron J., Badenhorst M., et al. "Weakness Associated with the Pathological Presence of Lipid in Skeletal Muscle: A Detailed Study of a Patient with Carnitine Deficiency. *Journal of Neurology, Neurosurgery and Psychiatry.* Vol. 39, 1976, pp. 1114–1123.

"Is Dietary Cholesterol Really Public Enemy #1?" *Women's Health Connection.* Vol. VI, Issue I.

Kagan, C. "Lysine Therapy for Herpes Simplex." *The Lancet.* Vol. 1, No. 37, 1974.

Lacour, B. DiGiulio S. Chanard J., et al. "Carnitine improves lipid anomalies in hemodialysis patient." *Lancet.* October 11, 1978, pp. 763–764.

Leibovitz, Brian. *Carnitine, The Bt Phenomenon. 1984.*

Lemon, P.W.R. and J.P. *Journal of Applied Physiology.* 1980, pp. 624–629.

Li, J.B., Jefferson, L.S. *Biocehmical Biophysical Acta.* 1978, pp. 351–359.

Milne, Robert, et. Al. *Definitive Guide to Headaches.* Tiburon, CA: Future Medicine Publishing, 1998.

Moore, Thomas. *Prescription for Disaster.* New York: Simon & Schuster, 1998.

Mowrey, Daniel. *Herbal Tonic Therapies.* New Canaan, CT: Keats Publishing, 1993

Moyers, Bill. *Healing and the Mind.* New York: Doubleday, 1993.

Murray, Michael T. *The Healing Power of Herbs.* Rocklin, CA: Prima Publishing, p. 208.

Murray, Michael T. "The Many Benefits of Carnitine." *The American Journal of Natural Medicine.* March, 1966., pp. 6–13.

Newbold, H.L. *Mega-Nutrients for Your Nerves.* New York: Peter Wyden, Publishing, 1975.

Neuborne, Ellen. "Workers in pain; employers up in arms." *USA Today*, January 9, 1997, Section B, p. 1–2.

Pert, Candace B. *Molecules of Emotion.* New York: Scribner Publishing, 1997.

Pfeiffer, Carl. Mental and Elemental Nutrition. New Canaan, CT: Keats Publishing, 1975.

Pfeiffer, Carl. *Nutrition and Mental Illness.* Rochester, VT: Healing Arts Press, 1987.

Poortmans, J.R., and In Howald H. *Metabolic Adaptation to Prolonged Physical Exercise.* Bsel, Switzerland: Birkhauseer Verlag, 1975, p. 212–228.

Prusinev, Stanley. *The Enzymes of Glutamine Metabolism.* New York, NY: Academic Press, 1973.

Rogers, L.L. "Glutamine in the Treatment of Alcoholism." *Quarterly Journal of Studies on Alcohol.* Vol. 18, No. 4., pp. 581–587, 1957.

Rogers, L.L. and R.B. Pelton. "Effect of Glutamine on IQ Scores of Mentally Deficient Children." *Texas Reports on Biology and Medicine.* Vol. 15, No. 1, pp. 84–90, 1957.

Sahley, Billie J. *The Anxiety Epidemic.* San Antonio, Texas: Pain & Stress Publications, 1997.

Sahley, Billie J. and Katherine M. Birkner. *Breaking Your Prescribed Addiction.* San Antonio, TX: Pain & Stress Publications, 1998.

Souba, Wiley W., et al. "Glutamine Metabolism by the Intestinal Tract." *Journal of Parenteral and Enteral Nutrition.* Vol. 9., No. 5 p. 608.

Shabert, Judy and Nancy Ehrlich. *The Ultimate Nutrient, Glutamine.* Garden City Park, NY: Avery Publishing, 1994.

Smith, Robert. "Chronic Headaches in Family Practice." *Journal of the American Board of Family Practice.* Vol. 6, Nov/Dec, 1992, pp. 589–599.

Thompsen J.H., Shug A.L, Yap V.U., et al: "Improved pacing tolerance of the ischemic human myocardium after administration of carnitine. *American Journal of Cardiology.* Vol. 43, 1979, pp. 300–305.

Tischler, M. et al. *Journal Biological Chemistry.* Vol 4, pp. 1613–1621.

"Riboflavin Zaps Migraines." *Vitamin Retailer.* May, 1998, pp. 53–54.

Whitaker, Julian. "Diabetes Treatment Review: How to Take Control," *Health and Healing Newsletter.* September, 1995, pp. 1–4.

Whitaker, Julian. *Reversing Diabetes.* New York: Warner Books, 1987.

Williams, R.J. *Alcoholism: The Nutritional Approach.* Austin, TX: Univ. of Texas Press, 1958, pp. 88, 100.

Zheng, J.J. and I.H. Rosenberg. "What is the Nutritional Status of the Elderly?" *Geriatrics.* June, 1989, pp. 57–64

Index

TO ORDER

Call 1-800-669-2256

Name ____

Address ____

City _______________________ State ________ Zip ____________

Books

The Anxiety Epidemic (Dr. Billie Sahley)	$9.95
Breaking Your Prescribed Addiction (Drs. B. Sahley & K. Birkner)	$9.95
Healing With Amino Acids (Drs. B. Sahley & K. Birkner)	$8.95
Malic Acid & Magnesium for Fibromyalgia/Chronic Pain (Dr. B. Sahley)	$3.95
Control Hyperactivity/A.D.D. Naturally (Dr. B. Sahley)	$8.95
Chronic Emotional Fatigue (Dr. B. Sahley)	$3.95
The Melatonin Report (Dr. B. Sahley)	$3.95
Is Ritalin Necessary? The Ritalin Report (Dr. B. Sahley)	$5.00
GABA, The Anxiety Amino Acid (Dr. B. Sahley)	$4.95
Breaking the Sugar Habit Cookbook (Dr. K. Birkner)	$8.95

AudioTapes by Dr. Billie J. Sahley $9.95 + P & H

Is Ritalin Necessary? The Ritalin Report Audiobook
Anxiety/Panic Attacks, Causes and Control
Hyperactivity, Causes & Control
Anger
Anxiety
Being, Your Way
Communication
Depression
Escape
Fear
Forgiving and Healing
Guilt
Letting Go
Phobias

SubTotal ________

(With Purchase Free catalog) Catalog $4 each ________

TX Residents Add 7.75% Sales Tax ________

U.S. Shipping** $3 (first item + $1 subsequent) ________

Note: Personal checks are held 10 working days. To expedite order, send money order.

MC/Visa/Discover _ _ _ _ - _ _ _ _ - _ _ _ _ - _ _ _ _

Signature ______________________________ Exp Date __/__

Send To: Pain & Stress Center, 5282 Medical Drive #160, San Antonio, TX 78229-6023

**Canadian and other foreign countries ADD $8 to the above amounts. We accept World Money Orders, or MC, Visa, or Discover cards ONLY!

About the Authors

Billie J. Sahley, Ph.D., is Executive Director of the Pain & Stress Center in San Antonio. She is a Board-Certified Medical Psychotherapist-Behavior Therapist, and an Orthomolecular Therapist. She is a Diplomate in the American Academy of Pain Management. Dr. Sahley is a graduate of the University of Texas, Clayton University School of Behavioral Medicine, and U.C.L.A. School of Integral Medicine. Additionally, she has studied advanced nutritional biochemistry, through Jeffrey Bland, Ph.D., Director of HealthComm. She is a member of the Huxley Foundation/Academy of Orthomolecular Medicine, Academy of Psychosomatic Medicine, and North American Nutrition and Preventive Medicine Association. In addition, she holds memberships in the Sports Medicine Foundation, American Association of Hypnotherapists, and American Mental Health Counselors Association. She also sits on the Scientific and Medical Advisory Board for Inter-Cal Corporation.

Dr. Sahley wrote: *The Anxiety Epidemic*; *Control Hyperactivity/A.D.D. Naturally; Chronic Emotional Fatigue; Malic Acid and Magnesium For Fibromyalgia and Chronic Pain Syndrome; The Melatonin Report; Is Ritalin Necessary? The Ritalin Report;* and has recorded numerous audio cassette tapes. She coauthored *Breaking Your Prescribed Addiction.*

In addition, Dr. Sahley holds three U.S. patents for: SAF, Calms Kids (SAF For Kids), and Anxiety Control 24.

Kathy Birkner is a Pain Therapist at the Pain & Stress Center in San Antonio. She is a Registered Nurse, Certified Registered Nurse Anesthetist, Advanced Nurse Practitioner, and Orthomolecular Therapist. She is a Diplomate in the American Academy of Pain Management. She attended Brackenridge Hospital School of Nursing, University of Texas at Austin, Southwest Missouri School of Anesthesia, Southwest Missouri State University, and Clayton University. She holds degrees in nursing, nutrition, and behavior therapy. Dr. Birkner has done graduate studies through the Center for Integral Medicine and U.C.L.A. Medical School, under the direction of Dr. David Bresler. Additionally, she has studied advanced nutritional biochemistry, through Jeffrey Bland, Ph.D, Director of HealthComm. She is a member of American Association of Nurse Anesthetists, Texas Association of Nurse Anesthetists, American Association of Pain Management, American College of Osteopathic Pain Management and Sclerotherapy, American Holistic Nurse Association, and American Association of Counseling and Development. She is author of *Breaking Your Sugar Habit Cookbook* and coauthor with Dr. Sahley of the audio cassette tape, *Therapeutic Uses of Amino Acids,* and the book, *Breaking Your Prescribed Addiction.*